Planning, Performing, and Controlling Projects

Principles and Applications

Third Edition

Robert B. Angus

Norman A. Gundersen

Thomas P. Cullinane

Prentice Hall

Upper Saddle River, New Jersey
Columbus, Ohio

Library of Congress Cataloging-in-Publication Data

Angus, Robert B. (Robert Brownell)
  Planning, performing, and controlling projects : principles and applications / Robert B.
  Angus, Norman A. Gundersen, Thomas P. Cullinane.—3rd ed.
    p. cm.
  Includes bibliographical references and index.
  ISBN 0-13-041670-3 (case)
  1. Project management. 2. System analysis. I. Gundersen, Norman A. II. Cullinane,
  Thomas P. III. Title.
T56.8.A53 2003
658.4'04—dc21   2002020082

Editor in Chief: Stephen Helba
Executive Editor: Debbie Yarnell
Editorial Assistant: Sam Goffinet
Media Development Editor: Michelle Churma
Production Editor: Louise N. Sette
Design Coordinator: Diane Ernsberger
Cover Designer: Jason Moore
Production Manager: Brian Fox
Marketing Manager: Jimmy Stephens

This book was set in Swiss 721, Dutch 801, and Zapf Humanist by TechBooks. It was printed and bound by R.R. Donnelley & Sons Company. The cover was printed by The Lehigh Press, Inc.

Earlier versions, entitled *Planning, Performing and Controlling Projects,* © 1995, 1993, 1992 by Bowen's Publishing.

Pearson Education Ltd.
Pearson Education Australia Pty. Limited
Pearson Education Singapore Pte. Ltd.
Pearson Education North Asia Ltd.
Pearson Education Canada, Ltd.
Pearson Educación de Mexico, S. A. de C. V.
Pearson Education—Japan
Pearson Education Malaysia Pte. Ltd.
Pearson Education, *Upper Saddle River, New Jersey*

10 9 8 7 6 5 4 3 2 1

ISBN: 0-13-041670-3

We dedicate this book to

Thomas E. Hulbert

and

Donald B. Devoe

who have worked with us so this book will be a valuable contribution to the education of science, engineering, and technology students

Preface

This text is an introduction to the theory and practice involved in the design and management of technically oriented projects. It is a merger of our personal experiences as designers, engineers, and technical managers of projects whose values have ranged from several thousand to a few million dollars. We have assumed that you are

- A student studying science, engineering, or technology
- Interested in the planning, performing, and monitoring of projects
- Expecting to become involved in projects that will grow in size and complexity

Other texts discuss project design and technical management. However, many of these texts are theory oriented and lack practical examples. (Texts that we feel would be useful to you as references are noted in the bibliography.) We have tried to achieve a balance between theory and practice by including a major case study that is introduced in chapter 1 and continues through chapter 5.

Our years of experience have taught us that the best theoretical knowledge is of little value until you apply it to practical problems. These problems include

- Estimating costs and schedules accurately at the beginning of a project
- Determining and fulfilling the expectations of the client
- Imparting theoretical and practical knowledge to new employees
- Guiding diverse groups of people toward a common goal
- Earning a profit so other projects can be pursued
- Enjoying the work as it progresses

What is the value of life experiences if they do not include (1) learning, (2) having fun while learning, and (3) earning the trust and respect of your superiors, fellow employees, subcontractors, and clients? Not much, we say!

The Northeastern University Cooperative Education Department's motto is "Learn while you earn." We add, "Enjoy your work, enjoy your employ," and prepare for your next opportunity. Test your new skills outside the classroom and practice the theoretical principles you have learned.

The philosophical basis of this book is the systematic approach to project design and management. This approach requires continually considering and reevaluating how each portion of a project interacts with the entire project. Its use can lead to an efficient, useful, and cost-effective product, process, or service.

A typical project using the systematic approach consists of four phases:

1. The Conception Phase, in which ideas are devised and brainstormed
2. The Study Phase, in which potential designs are investigated
3. The Design Phase, in which the system is actually designed in detail
4. The Implementation Phase, in which the system is constructed and delivered

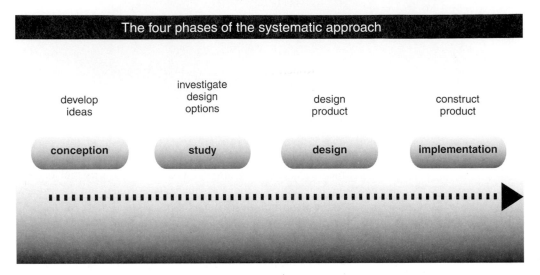

The four phases of the systematic approach

develop ideas — conception

investigate design options — study

design product — design

construct product — implementation

These phases may overlap. However, for discussion purposes, we consider them to be separate in time and cover them in separate chapters.

Chapters 2, 3, 4, and 5 each deal with one of these phases. The four chapters have the same organizational structure, with the following sections:

1. Purpose and goal of the phase
2. Activities occurring during the phase
3. Documentation for the phase
4. Defining the completion of the phase
5. Exercises

The example we use for the major case study is the result of our experiences in working with local business, industry, government agencies, and charitable foundations. In this example, Northeastern University engineering and technology students become involved in a community project in the town of Bedford, Massachusetts. The names of students used in the case study are fictitious.

Chapters 6, 7, 8, and 9 are specialty chapters. They describe our approach to systematic management, documentation, documentation interaction, and modeling. These chapters are an introduction to topics that may be of value to you in later years.

Colleges and universities using this book may need to adapt the sequence in which information is presented so that it dovetails with a variety of courses. The topics covered in the appendixes allow for the flexibility in presentation. The topics are

Appendix A	Value Analysis and Engineering
Appendix B	Objectives, Task Descriptions, and Active Verbs
Appendix C	Specification for the Town of Bedford: "An Activity Center"
Appendix D	Bedford Activity Center Contract
Appendix E	Bedford Activity Center Contract Change Notice (CCN)
Appendix F	Records and Their Interactions
Appendix G	System-Level Specifications
Appendix H	Student Activity Center As-Built Plans

Note that appendixes C, D, E, and H relate specifically to the major case study, the activity center for the town of Bedford.

A free full-product 120-day evaluation version of Microsoft Project 2000 software is provided on a CD-ROM at the back of the text. This CD-ROM allows users to apply the project management and documentation concepts presented in the text using this powerful project management software tool. Sample printouts from Microsoft Project are included in chapter 7.

Acknowledgments

We thank the following who have contributed to the third edition:

Michelle Churma, who guided us and shepherded all the changes

Robert Mager, who reviewed Appendix A and suggested several constructive changes and, most especially, for his development of behavioral objectives for educational application

Dawn Hawkins, Art Institute of Phoenix; Joseph A. Phillips, DeVry Institute of Technology (OH); and Mr. Kyle B. Smith, Assoc. AIA, ICBO, the reviewers of the third edition

We wish to thank again the following persons who assisted us in the preparation of the second edition:

Marilyn Prudente, copy editor, who worked closely with us to improve the readability, especially where ESL students are concerned

Thomas E. Hulbert, who rewrote the creativity and brainstorming material and then wrote a new Appendix A

Judith W. Nowinski, who reviewed and edited the entire manuscript

Tom Bledsaw, ITT Technical Institute (Indianapolis, IN); Stephen Howe, Wentworth Tech (retired); David A. McDaniel, ITT Technical Institute (Norfolk, VA); and Duane Nehring, ITT Technical Institute (Earth City, MO), who reviewed the second edition manuscript

We also want to repeat our thanks to those who contributed to the first edition:

Marilyn H. Steinberg, Science Coordinator for the Northeastern University libraries, for her assistance in locating research material

Judith W. Nowinski, Technical Editor for Bowen's Publishing

Janet Battiste, copy editor

Mark D. Evans, Assistant Professor of Civil Engineering at Northeastern University, and Eric W. Hansberry, Associate Professor of Design Graphics at Northeastern University, who acted as reviewers during manuscript development

Sunand Bhattacharya, ITT Educational Services, Inc.; J. Tim Coppinger, Texas A&M University; Charles H. Gould, University of Maine; David Hanna, Ferris State University; Roger Killingsworth, Auburn University; Peggy A. Knock, Milwaukee School of Engineering; Steven M. Nesbit, Lafayette College; Hirak C. Patangia, University of Arkansas at Little Rock; David Wagner, Owens Community College, who reviewed the final manuscript

Cleveland Gilcreast, Visiting Professor at Merrimack College, who reviewed the management details

Sunand Bhattacharya, ITT Educational Services, Inc., who provided valuable assistance and suggestions for illustrations, design, and pedagogy

Stephen Helba, Editor in Chief at Prentice Hall

Beverly Cleathero and Bill D'Annolfo for their contributions, especially to the specifications appendix

Those family members who have been patient while we worked on this book

Barbara Coleman, who cheerfully assisted us in proofreading every word and diagram at least three times; we especially thank her for her patience and accuracy

The writing of a book is a difficult task that we have enjoyed—with their support.

Bob Angus
Norm Gundersen
Tom Cullinane

English may be a difficult or second language for you.
The words in this book have been carefully chosen to ease the reading.

Contents

chapter *1*

1. Systematic Approach

The Purpose and Goals of this Book

Projects, Programs, and People

Planning for Performance

Applying the Systematic Approach

2. Conception Phase

3. Study Phase

4. Design Phase

5. Implementation Phase

1. Why are planning, performance, and control important in the execution of projects?

2. What is a project, and what is a program?

3. How are projects managed and evaluated?

4. Why is teamwork important on projects?

5. What are the steps in the life cycle of a project?

6. What are the four phases of a project that uses a systematic approach?

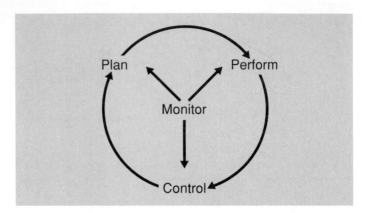

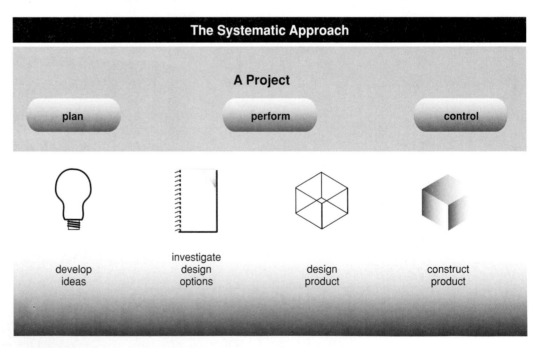

It is best to do things systematically, since we are only human, and disorder is our worst enemy.

Hesiod, 8th century B.C.

This book is intended to provide information on methods to plan, perform, and control a project.

Large and small projects require planning and control so they are completed

1. On time
2. Within budget
3. According to specifications

The achievement of these three goals enhances the reputation of the team, the manager, and the organization. Those who have successfully completed a project are in a better position to obtain new projects from any person or organization that wants quality work. In this book, we will discuss how projects can be initiated, planned, and controlled so that they are completed on time, within budget, and according to specifications. The specifications need to describe a product that is functional and constructable, as well as safe.

SECTION 1.1

*The purpose of this book is to present a systematic approach to the **planning, performance,** and **control** of projects.*

PLANNING AHEAD▶
■ ■ ■ ■ ■ ■ ■ ■ ■ ■ ■ ■ ■▶

You will see this symbol in many places within this text. It indicates some of the important areas where you should pause and plan ahead.

The case-study project will assist you in understanding the general concepts.

THE PURPOSE AND GOALS OF THIS BOOK

The purpose of this book is to present a systematic approach to the **planning, performance,** and **control** of projects. One goal of this book is to introduce you to techniques that will allow you to start, develop, and complete projects more efficiently and effectively. Another goal is to guide you in working as a member of a team. This is important because the global workplace requires people who are team workers.

Planning is the development of a method for converting ideas into a product or service. Scientists, engineers, and technologists learn to organize a collection of ideas into plans as part of their training. Projects are most successful when plans are carefully developed and prepared well in advance of design initiation (start).

Performance of a project requires a team of people working together to achieve the desired goal. Each team member needs to understand the plans and problems of other team members. As you study the material in the following chapters, you will become more aware of the relationship between your contribution and the successful performance of the entire project.

Control of a project includes seeing that it proceeds according to a well-prepared plan. However, we human beings do not devise perfect plans, since every possible circumstance cannot be known in advance, nor do we execute plans perfectly. Therefore, control includes recognizing when and why a project is not proceeding according to plan. Control also includes knowing if the plan needs to be revised or replaced with one that is more realistic, particularly if the client's expectations have changed.

Our approach in this book has three parts, as shown in figure 1.1:

1. Describe the general requirements for planning, performing, and controlling projects.
2. Explain how those requirements are applied through use of an example, a case study that we follow through several chapters.
3. Assign exercises that assist you in learning as you work on your own project.

The example project we have chosen is a combination of our personal experiences and those of others with whom we have worked. This project will assist you in understanding the more general concepts described in chapters 2 through 5 by means of a concrete example.

Figure 1.1 General approach of this book

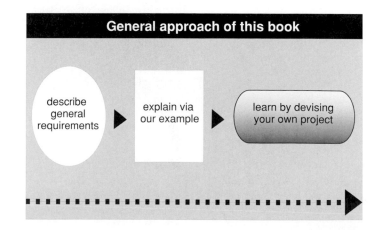

General approach of this book

describe general requirements ▶ explain via our example ▶ learn by devising your own project

SECTION 1.2

PROJECTS, PROGRAMS, AND PEOPLE

There are three reasons for initiating a project:

- *A bright idea*
- *A perceived need*
- *The need to improve a product*

What Is a Project?   A project is an organized effort to do something useful or attain a useful end result. It is sometimes referred to as a plan, venture, or enterprise. A project is usually guided by a leader known as the project manager.

There are three general reasons for initiating a project:

1. Someone has a bright idea worth being considered.
2. There is a perceived need to be fulfilled.
3. A product needs to be improved.

These reasons lead to the seven steps described in section 1.3.

Results may include such items as designs, plans, procedures, software, equipment, and structures. We later discuss how projects are initiated (conceived and started), planned (studied and designed), and accomplished (implemented and delivered).

How Do You Organize a Project Administratively?   To be successful, a project needs the following: (1) the project must be staffed with people who are able to direct it; (2) there must be sufficient funds to accomplish the project; and (3) there should be potential customers (clients and/or owners) to purchase and use the end product. A staff is required to plan, perform, and monitor the project. Experts are needed to provide advice. Performance of a project requires materials, equipment, and workspace (figure 1.2). These are *internal* influences. Administering a project is discussed at length in Fisk (1992). This and all other references are given in the bibliography.

Someone must be chosen to lead and guide a project. Committees can offer advice; however, they do not provide leadership. We will refer to the project leader as the **manager** of the effort. The manager's role is to organize **people, parts,** and **procedures.** The project manager must firmly establish the project's purpose and goal, as well as organize the group's resources. These resources include staff, ideas, experts, materials, workspace, and equipment. The manager must also focus on the needs of the customer and the funds available for the project.

Figure 1.2   The interrelated parts of a project

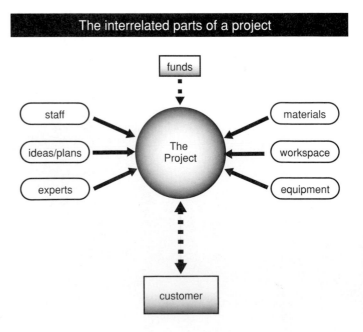

Figure 1.3 External influences that affect a project

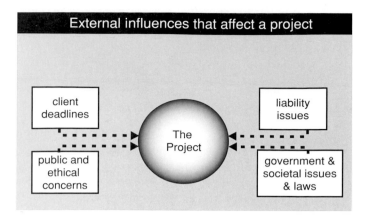

There are *external* influences that affect a project. Among them are the following:

- Client deadlines
- Liability issues
- Public and ethical concerns
- Government and societal issues and laws

It is the responsibility of the project team to continually weigh all these interrelated parts and external influences (figure 1.3) so that the final product is appropriate from both a technical and a social viewpoint.

A project is organized into portions that can be easily understood, accomplished, and controlled.

How Do You Organize a Project Technically?   A project is organized into portions that can be easily understood, accomplished, and controlled. For instance, a building to be constructed is organized as an overall shell having one or more floors. The plan for each floor is then designed separately. The floors require common subsystems, such as electrical power, air conditioning, and stairs, which are designed into an integrated system. A manager watches over the total system design, while technical specialists design the various common subsystems. Skilled drafters and designers lay out each floor and prepare a complete set of construction plans.

A project's success is judged by the project's design usefulness and its cost.

How Do You Evaluate a Project?   The performance evaluation of small projects is usually accomplished via performance reviews conducted by a senior staff member. The performance evaluation of large projects may require a separate staff of specialists who report to higher management. A project's success is judged by its design usefulness and its cost. Other factors involved in project evaluation will be discussed in this book in chapters 2, 3, 5, 6, 7, and 9.

What Is a Program?   A program is a group of related projects. A program usually requires a management team to oversee each of the projects and ensure that all the projects are coordinated. A program title might be "Devise and Implement a Housing Redevelopment Plan." That plan might consist of a group of projects, such as a revitalization of the community center, park improvements, and building painting. Each of these related projects is considered a separate project within the overall housing redevelopment program.

As a project or program becomes more complicated, the time required to accomplish it increases. Successful performance requires not only a plan, but also continual monitoring to verify that it is being followed or to identify when and where the plan must be modified as new information becomes available. For example, the assistance of additional people may be needed to complete a project. The effective coordination of people, parts, and procedures requires experienced management.

Figure 1.4 Types of skills needed for each phase

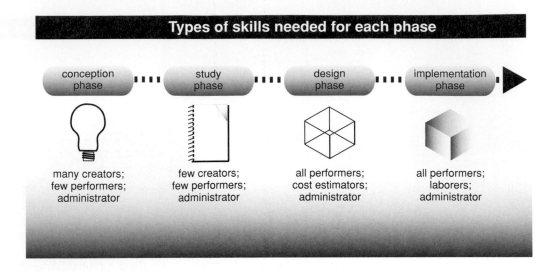

What Types of Skills Do Team Members Need?

The skills of team members should relate to the types of projects being planned and to the status of the project design. Examine figure 1.4, which is discussed in detail in chapter 6.

It is highly desirable for those involved in a project to understand technical subjects besides one's own specialty. Team members can then better evaluate the contributions of fellow team members whose skills may be different from theirs. Priest, Bodensteiner, and Muir (1994, p. 13) wrote that "[a] major study was conducted of over 1000 professionals from 22 high-technology companies to assess the educational and training needs of technical professionals engaged in the transition process from design to production. Eighty-six percent of the respondents believe that a technically based and multidisciplined engineering approach is fundamental to ensure a successful transition."

Another skill that is necessary is the ability of contributors to document and explain their work. Clear, concise, and timely documentation allows individuals' work to be evaluated, revised, and continued by other team members. For larger projects, the documentation is usually extensive. For some larger projects, the cost and quantity of documentation may be greater than the cost and size of the system being designed.

The project manager's skills must include the ability to (1) communicate with team members, clients, and other managers; and (2) evaluate and balance a variety of priorities as presented by others.

A project is devised, designed, managed, and performed by a group of people interacting with one another. These individuals are expected to place the needs of the team above their personal needs.

How Do Teams of People Work Together?

A project is devised, designed, managed, and performed by a group of people interacting with one another. These individuals are expected to place the needs of the team above their personal needs. This does not mean that the personal preferences and directions of each person should be ignored. On the contrary, each team member must learn to recognize that other people approach their work differently. All persons should respect the right of others to perform their assignments in different ways, as long as they contribute to the timely and successful completion of the project and do not interfere with the overall project effort. Each individual must realize that, when people work together, a better product will result, and it will contain the best ideas and efforts of each team member.

> For the many . . . when they meet together may very likely be better than the outstanding few, if regarded not individually but collectively . . . for each individual among the many have their share of virtue and prudence, and when they meet together, they become, in a manner, one [person], who has many feet and hands and senses . . . for some understand one part, and some another, and among them they understand the whole.
>
> Aristotle, 4th century B.C.

SECTION 1.3   PLANNING FOR PERFORMANCE

Steps in a Project

The seven steps of a project start with an idea and extend through sales and distribution:

1. Idea
2. Applied research
3. Design
4. Development
5. Marketing
6. Production (or construction)
7. Sales and distribution

This text is concerned with the first six steps. The steps for successful project performance are iterative rather than linear and are explained below. The sequence of steps may differ from one project to another.

Idea.   An idea is a concept first formulated in the mind of one or more persons. It may be a concept that can be designed or developed immediately, or it may require further study.

Applied Research.   Research, from "search again," is careful, patient, systematic, diligent, and organized inquiry or examination in some field of knowledge. It is undertaken to establish facts or principles. It is a laborious or continued search after truth or information. Applied research is an investigation whose goal is to convert an idea into a practical plan. If the applied research produces negative results, then the project will be terminated without further expenditure of funds and time.

Design.   Design refers to the conversion of an idea into a plan for a product, process, or procedure by arranging the parts of a system via drawings or computer models so that the parts will perform the desired functions when converted to an actual product.

Development.   Development refers to the activity of modifying a design, product, process, or procedure. It may include new features, a different appearance, or only minor cosmetic changes as it progresses from an idea to an actual product, process, or procedure. Development starts with a design and converts it into a product, process, or procedure. There may be several stages of experimental models, breadboard models, experimental prototypes, and production prototypes. The redesigned product, process, or procedure is now ready for production.

Marketing.   Marketing (or market research) involves determining the need for a new product, process, or procedure, and the amount of money that potential customers are willing to invest in that new product, process, or procedure. It is from a marketing study that the number of potential sales is estimated. The focus of the marketing effort is to determine if the new product will sell in sufficient quantities to meet the total project costs, as well as earn a profit. Marketing may occur prior to or during the design and development efforts.

Production.   Production is the act of making numerous, identical copies of an item that has been designed and developed. Production is more efficient and less costly when specialized manufacturing equipment and processes are used to produce the quantities required. However, the quantities to be produced must be great enough to justify the expense of specialized equipment and processes.

The life cycle flow

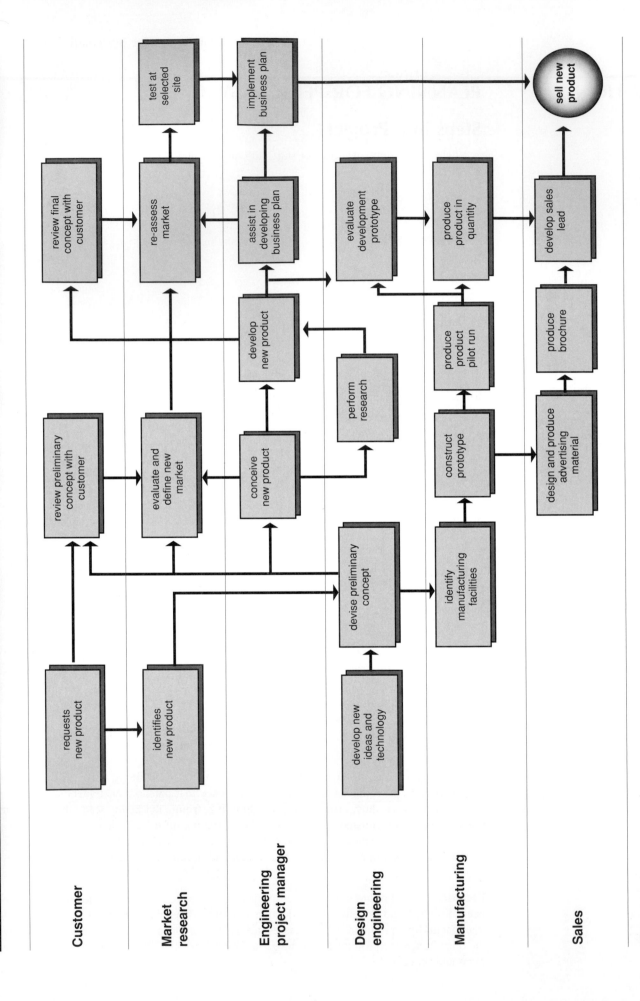

Customer

Market research

Engineering project manager

Design engineering

Manufacturing

Sales

requests new product

identifies new product

review preliminary concept with customer

review final concept with customer

evaluate and define new market

re-assess market

conceive new product

develop new product

assist in developing business plan

test at selected site

implement business plan

perform research

evaluate development prototype

produce product in quantity

develop new ideas and technology

devise preliminary concept

identify manufacturing facilities

construct prototype

produce product pilot run

design and produce advertising material

produce brochure

develop sales lead

sell new product

Figure 1.5   The life cycle flow

Sales and Distribution.   Selling is required if a project is to be financially profitable. For items manufactured in small quantities, sales consists of contacting those select customers who are willing to pay for the item. For items manufactured in large quantities, many customers are involved. Items often are packaged and sent to distributors, who then supply the products to retail outlets for final sale of the item to customers.

The Life Cycle

The life cycle of the project plan is complete when the initial idea has been converted, via applied research, design, development, marketing, production, and sales and distribution, into a successful and profitable product.

The life cycle of the project plan is now complete. The initial idea has been converted, via applied research, design, development, marketing, production, and sales and distribution, into a successful and profitable product. To accomplish this result, effective planning is necessary. The life cycle flow is shown in figure 1.5.

Concurrent Engineering

Concurrent engineering is a systematic approach that integrates the initial idea for a project with its applied research, design, development, marketing, production, sales, and distribution. Concurrent engineering leads to a more efficient and shorter design cycle while maintaining or improving product quality and reliability. See Banios (1992). Incorporated are:

- customer requests and ideas,
- market research regarding the potential product,
- design and development of models,
- manufacturing prototypes and a pilot run, and
- sales-material preparation and testing.

See figure 1.5 for an example of the overall cycle.

Concurrent engineering compresses the schedule; thus, many activities can be performed either simultaneously or with some overlap. The result is a shortening of the overall cycle time without adversely affecting the project quality, material delivery, and customer sales.

Schedules are devised that indicate when the normally sequential steps and phases of a project can be performed concurrently—that is, overlapped. Planning is based upon interim and final project goals; methods are then devised to achieve these goals. Next, the results are coordinated, and priorities are assigned and tied to customer needs. Engineering, manufacturing, and marketing must work together from the outset to anticipate problems and bottlenecks and to eliminate them promptly. However, no matter how streamlined the schedule, it needs to be remembered that a design actually desired by potential customers succeeds much better in the marketplace than a design that is merely acceptable.

Concurrent engineering recognizes the roles of marketing, manufacturing, and servicing in bringing products to market and in supporting them throughout their life cycle. It emphasizes the need for shared understanding of problems to be solved—as well as meeting the customer expectations as they evolve.

By using electronic communications networks, engineers have greatly reduced the time spent in reviewing designs with remotely located manufacturers. Technology allows design changes to be reviewed and implemented in a quick and systematic manner.

SECTION 1.4   APPLYING THE SYSTEMATIC APPROACH

This section describes in more detail the **systematic approach** to projects and programs first documented by Goode and Machol (1957). When applied wisely to both design and management, this approach most often leads to a less costly and more acceptable, efficient, and effective product or process.

*The life cycle of a typical **system** consists of research, conception, engineering, production, operation, and, finally, retirement or decommissioning.*

Applying the system concept requires maintaining an objective view of how each portion of a project interacts, or fits in, with the entire project. The life cycle of a typical **system** consists of research, conception, engineering, production, operation, and, finally, retirement or decommissioning. These steps are described in more detail in *An Engineering Practices Course* by E. W. Banios (1992). Our presentation focuses upon the conception, engineering, and production portion of the life cycle.

Definitions

As you progress through this book, you will encounter new words and phrases, as well as many words that are familiar to you from other contexts but that are used in the book in a specialized, or technical, way. To aid your understanding, here is a list of technical definitions of basic terms used in this book. These words have special meanings for scientists, engineers, and technologists.

Our Definitions

A **function** is the capability of an individual or equipment to perform in a given manner.

A **system** is any combination of things that actively work together to serve a common purpose. Systems are composed of combinations of parts, people, and procedures.

A **subsystem** is any portion of a system that performs one or more functions that can, alone or in a system, operate separately.

A **device** is that portion of a subsystem that can, alone or in that subsystem, perform an identifiable function.

A **component** or **part** is a portion of a device that normally performs a useful function only when used within that device.

Definitions of all other terms are given in the glossary.

Figure 1.6 shows how aircraft components, devices, and subsystems relate. The aircraft is the primary system for the aircraft designers. Three of its subsystems consist of the airframe, the internal electronics, and the people involved. The **airframe** subsystem consists of the body, wings, tail, and engines. The **electronics** subsystem consists of communications, navigation, and control devices. The **people** subsystem consists of the crew, passengers, and support personnel.

Here is another example from the aviation industry. The Federal Aviation Authority considers its traffic management system to be the primary system. Each participating airline and airport is a subsystem to be managed. Thus, at this level of consideration, each aircraft is a device to be guided, whether the aircraft is on the ground taxiing, taking off, in the air, or landing.

Figure 1.6   Component interrelation

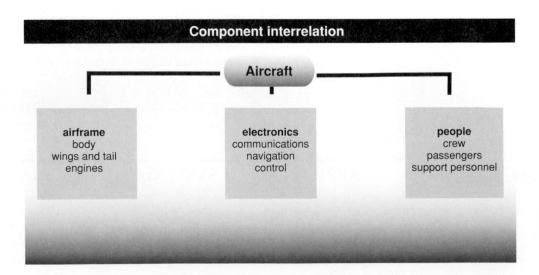

Phases of Project

A typical project or program that utilizes the systematic approach consists of four phases.

A typical project or program that uses the systematic approach consists of four phases between the consideration of an original idea and its delivery as a product or service. The four phases with their general descriptions are shown in figure 1.7 and are explained below.

Conception Phase: Ideas are developed and evaluated. Such practical considerations as technical feasibility and sources of funding need to be kept in mind.

The definition of an activity is given following the listing of the activities for the four phases. For the Conception Phase, they are:

- Organize a Team
- Formulate a Concept and Select the Project
- Initiate Conception Phase Documents
- Expand Team and Identify Supporting Organizations
- Complete the Planning of the Next Phases of Work
- Convert Document Drafts into Final Documents

Study Phase: Ideas developed in the conception phase are studied. One idea (or solution or direction) is selected and becomes the chosen system. Potential designs are investigated. Where necessary, applied research is performed.

For the Study Phase, the activities are:

- Organize the Study Phase Work
- Establish Ground Rules for Study and Design Phases
- Study Solutions to be Considered for Design
- Gather and Evaluate Information
- Select and Plan One Solution to the Project
- Document the Selected Solution
- Verify and Obtain Funding

Design Phase: The chosen system is designed in detail so that it can be constructed. Parts of the system design may need to be modified. More detailed cost information is available. Marketing can now establish a price and determine if the product will sell at that price.

Figure 1.7 The four phases of the systematic approach

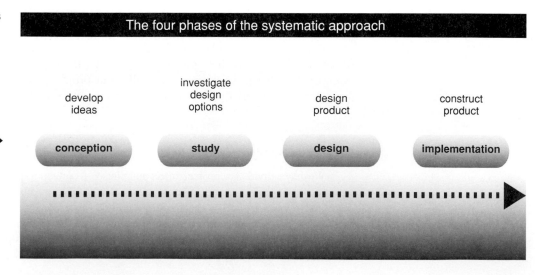

For the Design Phase, the activities are:

- Organize the Design Phase Work
- Select the Solution Details to be Implemented
- Document the Design Solution
- Prepare for the Implementation Phase Work
- Obtain Implementation Phase Funding

Implementation Phase: The chosen system is constructed and delivered.

Implementation Phase Activities are:

- Obtain Implementation Phase Funds
- Review Design Phase Documents
- Allocate Tasks to the Implementation Phase Team
- Prepare to Produce the Designed Article
- Produce and Evaluate Product
- Revise, Review, and Deliver Final Documentation

and, where applicable

- Operate and Maintain Customer-Contracted Articles
- Review the Results of the Implementation Phase
- Pursue New Projects

When a new project consists simply of a slight modification of a previously completed project, the Conception Phase will require less time and attention. Also, the Study Phase can either be shortened or merged with the Design Phase.

The four phases of a project that uses the systematic approach—Conception, Study, Design, Implementation—may overlap.

In chapters 2 through 5, the activities associated with each phase are described in detail. Even if a project is a small one, it will involve all or most of the activities described in these chapters. These phases may overlap. For discussion purposes, we will consider them to be separate in time, as mutually exclusive phases.

In chapters 2 through 5, we will discuss the purpose and goal for each phase. First, let us examine the definitions of the words *purpose* and *goal*.

Purpose: That which one proposes to accomplish or perform; the reason that you want to do something; an aim, intention, design, or plan.

Goal: An object of desire or intent; that which a person sets as an objective to be reached or accomplished; the end or final purpose; an ambition, aim, mark, objective, or target.

Note that there is some overlap between these two definitions. *Purpose* indicates the direction in which you want to proceed; *goal* indicates the end you want to achieve.

Before a project can be completed, it must be selected and started; the project must be **initiated.** We will discuss how projects are initiated in chapter 2 and include a case study as an example. In subsequent chapters, we will cover project planning, performance, and control—through a discussion of project phases and a project example.

Project Activities

We have divided a project into four phases—conception, study, design, and implementation. Each phase is subdivided into activities with one group of activities per phase. An **activity** is an active process, such as searching, learning, doing, or writing, that involves the application of both mental and physical energy. Each activity is further subdivided into smaller, more manageable steps or components until the deliverable products and documents are defined in enough detail to allow better management control.

The Case Study—Project Example

We have devised an example that will be used to illustrate the project phases, described in chapters 2, 3, 4, and 5. In the project example, first-year engineering and technology students apply their skills, capabilities, and interests to a project of their selection. We have selected Northeastern University in Boston as their school. The cities and towns described are suburbs of Boston, Massachusetts. All the student names are fictitious.

CHAPTER OBJECTIVES SUMMARY

Now that you have finished this chapter, you should be able to:

1. Explain the importance of planning, performance, and control in the execution of projects.
2. Define a project and a program; explain their similarities and differences.
3. Explain how projects are managed and evaluated.
4. Explain why teamwork is important on projects.
5. Name the steps in the life cycle of a project.
6. Describe the four phases of a project that uses a systematic approach.

EXERCISES

1.1   Choose a system that you have encountered during your daily activities. For that system, list some of its subsystems, devices, and components. Sketch a figure that presents the relationship among the subsystems, devices, and components. Consider what parts of your system could be considered a subsystem or device in another primary system.

1.2   Acquire a technical dictionary if you do not already have one. Compare the definitions given in section 1.4 with those given in your technical dictionary. Discuss with other students how and why they differ from the ones we have offered in this book.

1.3   List at least five projects and programs that you have either encountered or read about recently. For those projects and programs, write a brief paragraph on why they are labeled *project* or *program*.

Projects and programs might involve one or more of the following engineering and science specialties:

architectural, chemical, civil, computer (hardware and/or software), electrical, electronic, environmental, industrial, information systems, marine science, mechanical, systems engineering, and transportation.

1.4   Prepare a list of your personal interests and skills that would apply to working as a member of a team on a class project. Describe your previous experiences in performing and documenting work that you have accomplished. Consider at least the following questions:

Do you enjoy writing and sketching?

Are you able to devise and document original material?

Do you prefer to edit the writing of others?

1.5   With your instructor or faculty advisor, organize one or more teams to search for one or more class projects. Each project should require the application of more than one specialty. In the search for a project, ask, "Who will verify that all of the designs and activities are accomplished satisfactorily?"

chapter **2**

1. Systematic Approach

2. Conception Phase

Conception Phase Purpose and Goal

Conception Phase Activities

Conception Phase Documents

Defining Phase Completion

3. Study Phase

4. Design Phase

5. Implementation Phase

1. What are the purpose and goal of the Conception Phase of a project?

2. What tasks are involved in the Conception Phase of a project, and what activities or steps are performed to complete these tasks?

3. What is the process of selecting a project? How is brainstorming involved in the process? What criteria can be used to evaluate the choice of projects?

4. How are teams for projects organized, utilized, and expanded?

5. What kinds of risks may be involved in projects, and how are alternative solutions found?

6. How are tasks, schedules, and budget constraints identified for a project?

7. What documents are prepared during the Conception Phase?

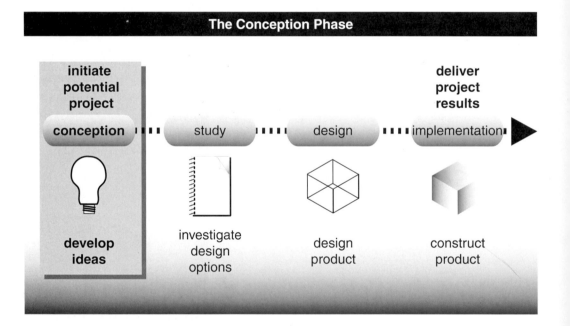

I have learned throughout my life as a composer chiefly through my mistakes and pursuits of false assumptions, not by my exposure to founts of wisdom and knowledge.

Igor Stravinsky, 1966

This chapter discusses how projects are first selected, how teams are assembled, and how tasks, costs, and schedules are estimated so project funding can be pursued.

CONCEPTION

SECTION 2.1   CONCEPTION PHASE PURPOSE AND GOAL

The **purpose** of the Conception Phase is to develop ideas for projects that are of interest and value to you, to potential funding sources, and to your project team. The **goal** of the Conception Phase is the preparation of a list of practical ideas that will lead to a viable project.

At the end of the Conception Phase, you should know how to accomplish the following:

- Define a problem and the risks involved.
- Identify both your customers and potential funding sources.
- Improve your skills and abilities, as well as those of the team, to meet the needs of the project.
- Identify others who have the skills and abilities that you and your team lack.
- Prepare documentation for your proposed solutions.
- Devise a plan that includes the work you must perform.

The plan produced at the initiation of a project will determine the remainder of the Conception Phase, as well as the work to be accomplished during the Study, Design, and Implementation Phases. The associated costs for converting the plan to practical ideas must be included during the Conception Phase. If the project is not selected until the beginning of the Study Phase, then the approximate costs of all plans must be included here.

During this phase, you will continually revise all plans, including cost estimates, when you obtain significant new information or when you develop new ideas. If you are part of a profit-oriented organization, you must plan to earn money in excess of your expenditures so that you and your team will have the financial resources to identify, propose, and accomplish new projects at a later time.

The project team for the Conception Phase consists primarily of diverse creators.

The project team for the Conception Phase consists primarily of diverse creators (Kim, 1990). (See figure 1.4.) A few performers are added later to provide the necessary details. As a project progresses from one phase to another, the number of creators is decreased. They are replaced by planners, performers, and administrators who are experienced in (1) working with the ultimate users and (2) performing various aspects of the project life cycle. (The project life cycle was discussed in sections 1.3 and 1.4.)

Once the creators' work on a project is over, the manager who does not want to lose their expertise must provide different work for them. This way they will be available as creators for future projects. Thus, the organization chart showing the interaction among people is often a fluid one. It must be adapted to the work being performed at the moment.

SECTION 2.2   CONCEPTION PHASE ACTIVITIES

The activities for the Conception Phase consist of the following:

Activity 1: Organize a Team and Formulate a Concept

Activity 2: Select the Project

    2A: Identify customers with a need to be fulfilled.

    2B: Locate sources of funds to fulfill that need.

    2C: Brainstorm to select projects to be considered.

    2D: Identify risks for each potential project.

    2E: Select one project and gather more information.

    2F: Search for and document alternative solutions.

Activity 3: Initiate Conception Phase Documents

    3A: Prepare the Project Selection Rationale.

    3B: Prepare the Preliminary Specification.

    3C: Prepare the Project Description.

    3D: List task, schedule, and budget constraints.

Activity 4: Expand Team and Identify Supporting Organizations

Activity 5: Complete the Planning of the Next Phases of Work

Activity 6: Convert Document Drafts into Final Documents

The sequence of activities given below may change from one project to another. However, all of these activities must be completed during the Conception Phase.

Activity 1: Organize a Team and Formulate a Concept

The tasks of the Conception Phase are these:

- *Develop ideas for projects*
- *Prepare budget and schedule estimates*
- *Write task descriptions*
- *Organize a team*

During the Conception Phase, you develop ideas, prepare accompanying budget estimates, write approximate task descriptions, and devise approximate schedules for each of the projects under consideration. If the project is complicated, you may devise a model. See chapter 9.

You must also select and organize a project team. The team members should be compatible, with complementary skills and abilities. They must all contribute to and investigate the proposed ideas.

Some individuals prefer to work alone on an assignment. However, someone must monitor the work of these individuals so that it remains on schedule, within budget, and is performed to meet the technical requirements of the project. These "loners" should be encouraged to interchange ideas with the other team members.

Realize in advance that the members of your team may have to be changed. Personnel changes are often required to meet the changing needs of your proposed project so that the project remains viable. Changes may also have to occur if significant personality conflicts arise. Remember—a team that can work together *effectively* is being assembled, organized, and assessed.

One team member must be designated as team leader or project manager. This person has the responsibility of organizing and directing the efforts of all the other team members, as described in chapter 6. This manager (or leader) should be a decisive person with both technical and people skills. The manager directs, controls, and verifies project performance and status. The manager also adds new team members as the need for new skills and abilities is identified and terminates or transfers them to other projects when they are no longer needed.

Projects must be practical. They should

- *Meet the needs of the community*
- *Improve skills of team members*
- *Be capable of being funded*

The typical sequence for project conception is shown in figure 2.1. Your proposed projects should focus upon those ideas that

1. Consider the needs of the local community;
2. Use and improve your personal and team skills, abilities, and interests;
3. Concentrate on what you and your team can do well;
4. Identify projects that can be funded by interested people and organizations.

Figure 2.1   Project conception

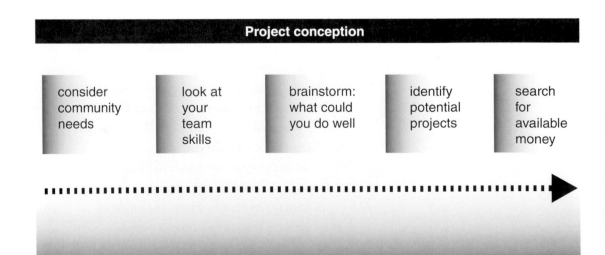

CONCEPTION

During this phase you will continue to gather applicable technical data that will assist you in arriving at a realistic concept while you prepare rough plans, and cost and schedule estimates. Later, you will examine how to use the brainstorming concept to expand upon your initial ideas for projects in a positive way. Brainstorming occurs before you select a specific project. Brainstorming is the most commonly applied of the many varied techniques for idea generation. Some of these techniques are described in appendix A.

While you consider and evaluate a variety of projects, you are actually developing ideas that might become the selected project. Expand all ideas and gather enough information for each potential project to allow you and your team to select one project to pursue in detail. Team members should refrain from criticizing evolving ideas during the brainstorming or any other idea-generation process. As you consider potential projects, you should also explore related projects and gather additional information associated with them.

PLANNING AHEAD▶

As you plan a project, your plans will include answers to questions that start with *who, what, when, where, why,* and *how* (see figure 2.2):

Who will be responsible for the work?

What is to be accomplished and by whom?

When is it to be implemented?

Where is it to be performed?

Why should it be performed?

How will the performance of the project be controlled?

Before you formally initiate a project, you must justify that project. Consider these questions:

Who would be the potential customer and is the proposed project worthwhile to the potential customer?

What will be the approximate cost of the project?

When does the customer require the project to be completed?

Where will the funds for the project originate?

Why should this project be pursued instead of another project?

How will be the project be accomplished?

You and your team must answer these questions, and perhaps other questions, before a significant amount of money and effort is expended.

Convene meetings during which various proposed projects are compared and the questions *who, what, when, where, why,* and *how* are asked. Often, if you combine several proposed projects,

Figure 2.2   The key questions in project planning

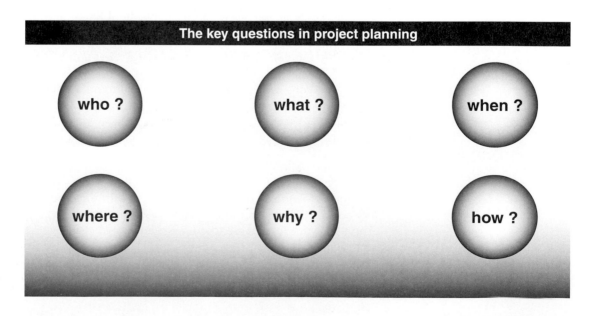

The key questions in project planning

who ?   what ?   when ?

where ?   why ?   how ?

the combined project will lead you to a better one. You should make such decisions only after you have gathered and compared enough information. (How much information you gather and compare requires a qualitative judgment.)

Teams must review all potential projects prior to extensive planning for any one project.

For each proposed project, you must prepare rough plans that provide details concerning the additional work required for you to perform the next three phases. Your plans should include the best available approximations to the required tasks, schedules, and costs. (These three interrelated documents are described in chapter 7.) While you prepare rough plans, and cost and schedule estimates, you will continue to gather applicable technical data to assist you in arriving at a realistic plan. Consider such factors as project cost and complexity, and identify potential customers. Your estimates will change as more data and ideas are gathered, but you must review all potential projects prior to extensive planning. Who will be helpful during such reviews? The manager must choose these people carefully.

The shaded portion of this text, starting below, presents a project example (major case study) that provides specific examples of the general activities described. If you desire to examine only the case study material, then read just the shaded portions of this and succeeding chapters. They flow together and provide a consecutive description of our project approach.

PROJECT EXAMPLE

Professor Thomas E. Hulbert has organized a team of six first-year engineering technology students from Northeastern University in Boston, Massachusetts. Under his guidance, they investigate several local cities and towns in search of a project to perform. Their participation in the project is designed to enhance their skills and abilities in the civil, electrical, mechanical, and industrial fields of engineering and technology.

The team must identify customers, locate sources of money, and select a project in the first six weeks.

Starting October 1, this team is allowed six weeks to perform the Conception Phase activities. They must first identify customers who have a need that could be fulfilled by a project. Then they must locate sources of money that could fund the project. Their selected project must be implemented within a twenty-six-week school term that begins on March 1 and continues through the end of August.

The initial team consists of students from a variety of backgrounds:

Roger Stein had a paper route; worked in a supermarket; is friendly; likes being around people.

Alice D'Annolfo worked with her father and uncle in construction; enjoys working outdoors.

Laura Drake was involved in high school band and field hockey; is creative and is fascinated by science and technology.

Jose Hernandez worked in a pizza parlor; is expressive; likes to work with his hands and with computers.

Kim Wong worked in her parents' restaurant; is friendly; has an interest in chemistry.

Ellery Kistler is a Russian immigrant with drafting experience; is shy and retiring; enjoys designing and constructing model ships.

After three ballots, the students' choice for team leader is Roger Stein.

Roger organizes the six of them into three teams of two persons. Their initial assignment is to discuss potential projects with local business leaders and members of Northeastern's Industry Advisory Committee for Engineering Technology. The teams are to provide interim written reports to Roger; these reports document each of their discussions. Then the group is to determine which potential projects could be performed within the given time. The students will meet formally with Professor Hulbert within three weeks and report the combined results of their activities.

After comparing their personal interests, the team members decide that a small construction or repair project could be accomplished within the time allotted and with the skills available.

Next, they consider the *who, what, when, where, why,* and *how* questions:

Who will be responsible for searching for a project? Roger assigns that responsibility to Alice because of her family background and her own personal interests.

What is to be accomplished and by whom? Roger will search for funding sources, with the guidance of Professor Hulbert. Ellery is asked to work with Laura in developing sketches of any ideas to be considered.

When is it to be implemented? The group agrees that the entire project must be accomplished within the remaining twenty-six weeks of the school year.

Where is it to be performed? The project must not require more than twenty miles of travel because there are only two persons with cars.

Why should it be performed? The project must use and improve the skills of the six team members. It should also apply the skills, abilities, and interests of other students in their first-year class if they become team members.

How will the performance of the project be controlled? Jose and Kim are assigned to develop the necessary tasks, schedules, and budgets. They may later become responsible for monitoring the project tasks, schedules, and budgets once these are defined. Roger will be responsible for ensuring that the quality of the finished product meets whatever specifications are agreed upon by the team members and the potential client.

The team members note that the Conception Phase has been allotted only six weeks. Thus, they decide to explore potential projects for the first two of those six weeks.

Three categories of nearby potential customers are established for the student teams: business, government, and industry—one for each of the three teams. Each team is to work through the contacts of their professors and establish potential projects.

They also decide, in advance, to concentrate on small renovation or construction projects, as these will best utilize their skills. Three categories of nearby potential customers are established: business, government, and industry—one for each of the three teams. Each team is to work through the contacts of their professors and establish potential projects.

Formal project initiation will occur after all the information has been gathered, and each potential project justified. The group prepares a list of questions—an informal questionnaire—and obtains appointments with local business, government, and industry officials through the university's Industrial Advisory Board. The questions include these:

Who would be the potential customer, and is the proposed project worthwhile to that customer?

What will be the approximate cost of the project?

When does the customer require the project to be completed?

Where will the funds for the project originate?

Why should this project be pursued instead of another project?

How will the project be accomplished?

Is the project composed of parts that will fit together smoothly?

Is there a definite sequence for optimum performance of the project?

Can the project be completed within a time acceptable to the customer?

Are the team members able to accomplish the necessary work within available funds?

Will the completed project improve team members' skills and abilities?

Is there a technical or financial risk in pursuing the proposed project?

On a weekly basis, Professor Hulbert meets with each team to assess its progress and guide it toward its final goal. He insists that the students prepare a trip report for each meeting, whether the visits are single or team visits. The team members feel apprehensive about interviewing persons unknown to them and about the amount of effort and writing they anticipate. Professor Hulbert indicates that this work is an important part of the Conception Phase initial research. He also informs them that they may feel more comfortable by first talking with and interviewing people they know personally regarding ideas for potential projects. Following these experiences, they can then visit and interview persons unknown to them and have a greater feeling of confidence.

On Monday afternoon of the third week, Roger convenes a meeting where the potential projects are compared and the answers to the questions *who, what, when, where, why,* and *how* are examined.

Each of the three teams is given one week to develop preliminary plans that provide details regarding the work that will be required for the next three phases of the potential project. The teams are instructed to develop approximations of the required tasks, schedules, and costs. Professor Hulbert invites three of the Industrial Advisory Board members to attend the review of the team efforts.

Activity 2: Select the Project

During this activity, you perform the following tasks:

1. Identify customers who indicate they have a need; establish the technical problems and risks associated with their need.
2. Determine if the potential customers have the funds for fulfilling that need and separate from the stated "need" their actual desires and luxuries.
3. Locate the source of those funds.
4. Brainstorm to determine what projects should be considered.
5. Identify the problems and risks associated with each project.
6. Search for acceptable alternative solutions to each proposed project.

The process of selecting a project is often known as *synthesis* (Pugh, 1993, p. 68). The synthesis process includes two major cyclical components:

- The generation of concepts that are solutions to meet the stated need.
- The evaluation of these concepts to select the one that is most suited to matching the stated need.

Concepts are often best generated by individuals; concept selection and enhancement is often best performed in groups. The outcome of the [Conception] Phase is a complete concept, engineered to an acceptable level to establish its practicality. Details are saved for the Study and Design Phases.

Activity 2A: Identify Customers with a Need to Be Fulfilled

Your projects must have a useful purpose. Therefore, as you evaluate potential projects, members of your team should determine the requirements of the local community and industry. These requirements are often categorized as needs, desires, or luxuries. If money is no object, then the distinctions between needs, desires, and luxuries are unimportant. If money is a significant factor, then the requirements must be separated into needs, desires, and luxuries.

What are the applicable definitions of needs, desires, and luxuries?

If money is a significant factor in developing a project, then project requirements must be separated into needs, desires, and luxuries.

Needs. A need is something that is considered necessary or essential. During World War II, Dr. Abraham Maslow studied the hierarchy of needs. He claimed that certain needs become dominant when more basic needs are satisfied. Thus, the lower-level needs must be satisfied before the higher-level needs become apparent. The list below indicates his hierarchy:

- Physiological needs: air, food, water
- Safety needs: protection against danger, threat, deprivation, arbitrary management decisions; the need for security in a dependent relationship
- Social needs: belonging, association, acceptance, giving and receiving of love and friendship
- Ego needs: self-respect and confidence, achievement, self-image, group image and reputation, status, recognition, appreciation
- Self-fulfillment needs: realizing self potential, self-development, creativity

Some persons or groups may believe that they have a need. The following questions may help you test if a need is real or imagined:

- Can you determine if the need is real or imagined?
- Are they or you able to express that need in writing?
- Can the need be separated into parts?
- Is each part important enough to continue to be considered as a separate need?
- How serious are you about committing funding to fulfill the need?

The persons or groups may wish to have their "need" fulfilled, yet they may not be able to justify the time and money to do so. In this case, the "need" may be merely a desire.

Desires. A desire is similar to a wish. It is a longing for something that promises enjoyment or satisfaction. A desire, by definition, is not a need. A strong desire may eventually become a need; a weak desire may be considered a luxury.

Luxuries. A luxury is the use and enjoyment of the best and most costly things which offer the most physical, intellectual, or emotional comfort, pleasure, and satisfaction. A luxury is considered to be unnecessary to life and health.

The interrelationship of needs, desires, and luxuries is illustrated in figure 2.3. The three definitions given are qualitative, not quantitative. One of your most difficult tasks during the

Figure 2.3 Comparing needs, desires, and luxuries

Comparing needs, desires, and luxuries

most **significant** least

needs	desires	luxuries
housing	purchase car	purchase sports car
food	purchase home	purchase 2nd home
clothing	join health club	visit France

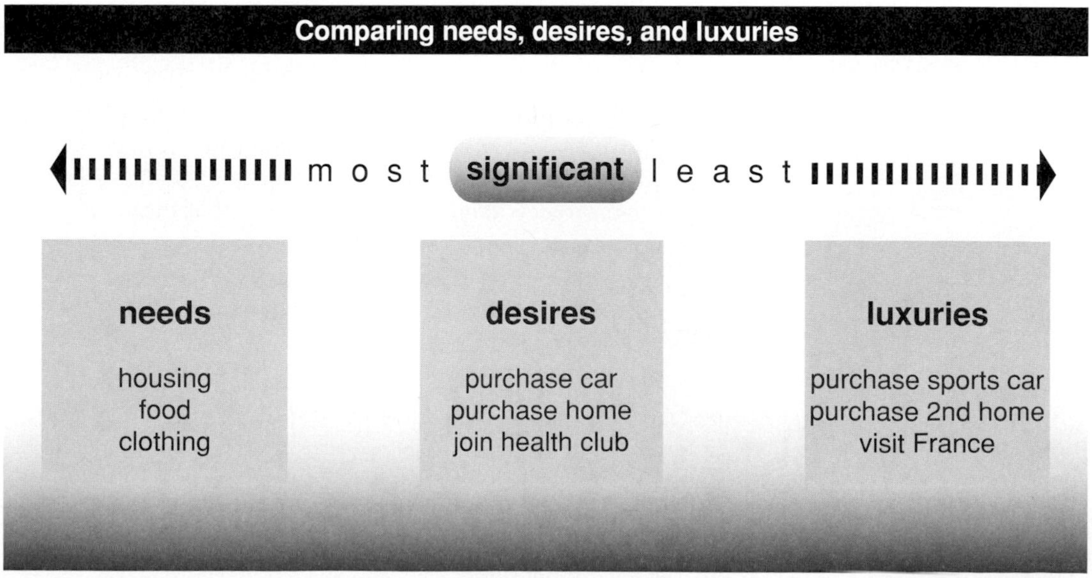

Conception Phase is to assign numerical values to needs, desires, and luxuries, and to weigh them with regard to the total system concept as described by von Oech (1993). However, if the customer (client) is willing to pay for it and the luxury is ethical, then the project can include the luxury.

Systems are normally implemented to provide for needs, rather than for desires and luxuries.

Systems are normally implemented to provide for needs. However, if you, the contractor, can include the desires and luxuries at minimal additional cost, you may be selected as the winning bidder. You must use judgment to determine the proper emphasis among needs, desires, and luxuries because it is unlikely that the potential customers will express their requirements in a clear and concise manner. The proposal evaluator, or team of evaluators of all competing proposals, will also be using judgment regarding the selection of the winning proposal. As Wilde (1978) notes: "As in life, instinct often precedes wisdom."

Market research is best initiated during the Conception Phase. People who specialize in market research have abilities that range from examining patents, trademarks, registered designs, copyrights, scientific journal proceedings, and reference books to less technical skills that include determining the need for a new product, process, or procedure—a new item—and the amount of money that potential customers are willing to invest in that new item. Because market researchers must often estimate the sales potential of a proposed item, they must try to determine if the new item will sell in sufficient quantities to meet the expected total project costs and earn a profit for the company. They must be involved in, or kept informed of, the total design status. However, they are often not trained in a background of total design and are thus unlikely to ask the right questions. See Pugh (1993, pp. 155–157). Some typical questions to ask potential customers or clients are the following:

- What performance do you require, and in what operating environment?
- What are you prepared to pay for this performance?
- How soon do you want the product, process, or procedure?
- What are your expectations regarding quality, reliability, life expectancy, and level of maintenance and documentation?
- What do you like best, or dislike most, about your competitor's designs?
- What safety factors and ergonomic features do you expect to be a part of the new product, process, or procedure, and how are they to interact with your related systems?

All design should start with a need that, when satisfied, will either fit into an existing market or create a market of its own. From the statement of that need, a Preliminary Specification is prepared during the Conception Phase that will evolve into the Design Specification in the Study Phase. Pugh (p. 104) provides a design sequence that starts with a market analysis, includes study, design, development, and production, and ends with selling the product.

PROJECT EXAMPLE

Alice discovers, through her uncle, that the planning board of the town of Bedford, Massachusetts, has devised a town redevelopment plan that is being circulated to contractors for review. Her uncle sends Alice a copy of the plan. There are five projects (figure 2.4) in the plan that the team might have the skills to implement:

- Pave walkways
- Improve playground
- Improve swimming pool
- Construct children's activity center
- Replace old carousel

The team meets and considers these five projects. They visit officials in Bedford to obtain additional information and to see if it is possible to separate needs, desires, and luxuries. They are told that *price* will greatly influence what Bedford can do and will determine the separation among needs, desires, and luxuries.

Town officials note that some town and state money is available, but it is not enough to accomplish any one project. The team must determine the availability of other money and its sources. The team decides to follow the path indicated in bold type in figure 2.4.

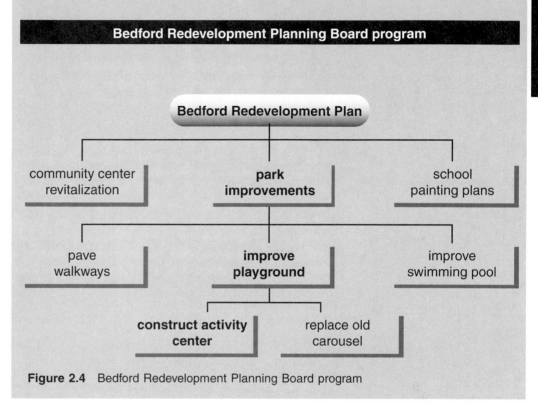

Figure 2.4   Bedford Redevelopment Planning Board program

Activity 2B: Locate Sources of Funds to Fulfill That Need

*There are three types of funding sources to be considered: **government, charitable organizations,** and **lending institutions.***

There are three types of funding sources to be considered: **government, charitable organizations,** and **lending institutions.** Money to fund the project may come from one or more local, state, or federal government agencies. It may be an investment from a bank or other funding or lending agency. It may also be a donation from a corporation or foundation.

When foundations **donate** funds to a worthy project, the foundation executives must believe that the image of their foundation will be enhanced in the minds of the general public when its donation to that project is announced. Organizations that donate money usually want to see the results of their donation within a preestablished length of time. They want their gift to provide something useful to the world as they envision it; they want to know that their donation was not wasted.

Institutions that **invest** their money want it repaid within an agreed time and with interest payments for its use—their **return on investment** (Heyne, 1991). The borrower may also be required to pay for a perceived risk, which would originate from the lender's worry that the borrower might not be able to repay the borrowed funds if the project should fail. In a case of perceived risk, some lenders may require that the borrower provide loan collateral, a security given as a pledge for the repayment of an obligation. This security may be in the form of bonds, stock certificates, or other property of value, and the loan would then be referred to as a "secured loan." Other lenders may rely on the reputation of the borrower, the credit rating of those involved, or the perceived merits of the project itself to guarantee repayment.

PROJECT EXAMPLE

Before the team selects a project, the following information is gathered:

The town of Bedford has only enough funds for the school painting plans and the swimming pool improvement. The selectmen (the town governing body) prefer to use the funding for new construction rather than the funding of repairs.

The Commonwealth of Massachusetts has $15 000 allotted to assist in projects that will provide summer work for young people. All the projects proposed by the selectmen cost more than the allotted $15 000.

An additional $5 000 was discovered to be available from a fund provided to the town via the will of Eleanor Kewer, a retired teacher. The money from this fund must be used for the construction and maintenance of facilities for school-age children.

Roger Stein approaches Professor Hulbert, who then calls both a selectman in Bedford and the attorney responsible for the commitment of the funds of Eleanor Kewer. The attorney encourages the team to work through Northeastern University, which would be responsible for the effort. Professor Hulbert sends to the selectman and the attorney a brochure that summarizes the types of community-oriented projects previously performed by other teams of students at Northeastern University. The selectman and attorney ask that Northeastern prepare a proposal for their consideration. They indicate a priority interest in either constructing a children's activity center or replacing the old carousel.

Normally you must acquire a small amount of funding from an initial funding source for all or most of the Conception Phase. If the labor of team members is your only significant initial requirement, then perhaps all the funding you need would be to pay for office supplies. These supplies might be borrowed. For example, you may be able to utilize computers on a time-available basis from other projects.

Authorization must be obtained from the funding source before money can be transferred from the source to a project.

At the same time that you are seeking funding for the Conception Phase, you are also identifying potential sources of funding for the entire project. Your project can begin when money is available for it. You must obtain authorization from your funding source before money can be transferred from the source to your project.

Activity 2C: Brainstorm to Select Projects to Be Considered

Brainstorming is the unrestrained offering of ideas and suggestions by all members of a team during a meeting that focuses on a predetermined range of potential projects.

The team must develop project options that are acceptable to all members. One technique for devising new ideas is **brainstorming,** defined as an *unrestrained* offering of ideas or suggestions by all members of a group, typically during a meeting that focuses upon a predetermined problem or a range of items.

Brainstorming involves **creativity,** which is an important part of the Conception Phase (Kim, 1990). We use the word "creativity" to indicate a new way of thinking or looking at a problem. When new ideas are being considered, they often lead to an invention or innovation. Albert Szent-Györgyi once said, "Discovery consists of seeing what everybody has seen, and thinking what nobody has thought." The invention or innovation may be a new product, process, or procedure, or it may be the improvement, revision, or redesign of an existing one.

Young children are naturally creative because their minds are relatively unrestrained. In adulthood, our range of thinking narrows because we have been trained to conform, "play it safe," and follow the rules. Roger von Oech, in his book *A Whack on the Side of the Head*

(1993), provides a list of common ways of thinking that block creativity and explains why these ways of thinking should be discarded when we are trying to generate new ideas:

1. Find the "right" answer: There's often more than one "right" answer.
2. That's not logical: Note that logic can eliminate alternatives that seem to be contradictory, but they may turn out to be good ones.
3. Follow the rules: Change or break those rules that are no longer sensible.
4. Be practical: Ideas initially need the wide realm of the possible rather than the narrow realm of the practical. Think: "what if . . . ?" because useful ideas often occur to people while considering many useless ideas.
5. Don't be foolish: Behavior that is silly or humorous often leads to new ideas.
6. Don't make mistakes: Mistakes can be stepping stones in the creative process.
7. That's not my area: This attitude often keeps people from trying new paths. You already know what cannot be done and this can restrict the creative process. New ideas often originate when you venture outside your field of specialization.
8. I'm not creative: Self-esteem is essential to being creative; you must be willing to take risks and accept the consequences, which may be ridiculed by less creative people. If you believe that you are creative, or can be creative, then you will be creative!

Most adults need to be stimulated into new thinking; it seldom occurs spontaneously. We need to perform exercises to remove the mental blocks to new thinking. Here are some techniques you can use to help yourself to become more creative:

1. Stimulate your thinking by reversing the sequence in which you put on your clothes, such as shoes, underwear, shirts, sweaters. Reverse the right and left sequence for inserting hands and feet; it will stimulate your rusty brain cells.
2. Imagine (or observe) how a preschool child thinks about something. Children are less limited in their creativity because their school peers have yet to tease them and discourage their creativity.
3. Record your ideas as soon as you think of them. For instance, have a pad and pencil beside your bed and document your dreams.
4. Eliminate the mental blocks that prevent you from being creative. Focus upon positive thoughts. There is an old song that starts, "Accentuate the positive; eliminate the negative. . . ."
5. Prepare questions that will stimulate a variety of answers. Remember: Answers depend upon the questions asked.
6. Read something outside your area of specialty, preferably something different that you will enjoy, such as fiction or poems.
7. Laugh at yourself. A sense of humor helps because acting silly can lead to new ideas.

For your team project, initiate a nonjudgmental brainstorming session to develop a list of potential projects. The session can be structured or unstructured. A structured session requires that each person in the group be solicited in sequence and provide an idea each time, or "pass" until the next time. An unstructured session allows group members to offer ideas spontaneously. In order for the session to be successful, the team must agree to follow certain rules that have been found to be effective in promoting the uninhibited flow of ideas. In his book *Applied Imagination* (1993), Osborn recommends that brainstorming groups follow these four basic rules:

1. Defer all criticism of ideas: Participants must withhold negative judgments about suggestions. Osborn writes, "When driving for ideas, you go farther if you keep your foot off the brake."
2. Encourage a "free-wheeling" approach: The wilder the idea, the better; it is easier to tame down than to think up.
3. Go for quantity: The greater the number of ideas, the more likely there will be some winners in the bunch.
4. Combine and improve on others' suggestions: In addition to contributing ideas of their own, participants should try to add their own ideas to the suggestions of others.

In addition to Osborn's four rules, the authors and other practitioners experienced in organizing brainstorming sessions recommend the following guidelines published in the ASEE *Prism* magazine (March, 1998). These techniques have proven effective when applied to both large and small projects:

1. Carefully define the problem at the start of the brainstorming session to avoid wasting time on the wrong problem.

2. Allow individuals five minutes to consider the problem before the group tackles it to prevent a "follow-the-leader" thought process.

3. Create a comfortable, casual environment where participants do not feel intimidated so they won't speak up. A circle arrangement is a good setup. If a non-circular table is used, have the "leaders" sit in the middle of a long side to reduce any intimidation that may occur from sitting at the "head" of the table.

4. Record all suggestions. Suggestions may be lost otherwise, and documenting each one lets the participants know that the group accepts and acknowledges every idea.

5. Appoint someone, a group member or another individual, to serve as a facilitator. This person ensures that all members are following the rules.

6. Encourage participants to combine and piggyback if idea generation lags after a session starts. Ideas begin as traditional and tend to become more innovative later in the meeting.

7. Stop the session before the group fades on idea generation. Identifying the best stopping time requires practice and experience on the part of the leader.

8. Keep brainstorming groups small, from six to twelve members. It is also helpful to have participants whose backgrounds range from total familiarity with the problem to total separation from the problem.

9. After the session, classify the solutions using ABC: *A* for feasible solutions, *B* for ideas requiring further development, and *C* for long-term development and technology.

10. Report results to participants and thank them for their time and effort.

Checklists help brainstorming participants develop a budding idea or alter an old one. Run through a checklist of ways to change a current product or process. Osborn has developed the following checklist for change:

Adapt: What else is like this? What other idea does this suggest? Does the past offer a parallel? What could I copy? Whom could I emulate?

Modify: Could I add a new twist? Could I change the meaning, color, motion, sound, odor, form, or shape? What other changes could I make?

Magnify: Could I add something, such as more time or greater frequency? Could I make it stronger, higher, longer, or thicker? Could I add more ingredients? Could I duplicate, multiply, or exaggerate it?

Minify: Could I subtract something? Could I make a condensed or miniature version? Could I make it lower, smaller, shorter, or lighter? What if I omit something or streamline, divide, or understate it?

Put it to other uses: Are there new ways to use this as it is? Does it have other uses if I modify it?

Substitute: What if I used a different ingredient, material, person, process, power source, place, approach, or tone of voice?

Rearrange: Could I interchange components? Could I use a different pattern, layout, or sequence? What if I transpose the cause and effect? What if I change the pace or the schedule?

Reverse: Could I transpose positive and negative? What if I tried the opposite or reversed roles? What if I turn it backwards, upside down, or inside out?

Combine: Could I use a blend, an alloy, an assortment, or a combination? Could I combine units, purposes, appeals, or ideas?

Brainstorming can be performed at two levels. The first level concerns brainstorming ideas that could become projects. All ideas are documented and studied later so the better ones can be pursued in more detail. Once a potential project is selected, brainstorming can be applied to devising clever solutions for the selected project (see chapter 3). This second level concerns developing the details within the selected project (figure 2.5). The brainstorming approach is the same for both levels.

The following examples of projects require different sets of skills. Choosing one of these projects represents the first level of brainstorming.

- Design land usage for a housing project
- Design and construct a community center
- Design and build a model railroad control system
- Design and construct a solar-powered car
- Develop equipment for handling hazardous waste materials
- Develop procedures for handling recyclable materials
- Write new software program for . . .
- Develop a better procedure for . . .

On completion of the brainstorming session for these projects, each idea must be evaluated. Now reality must be added to the creative effort resulting from the brainstorming session.

Developing a list of project selection criteria such as the following represents the second level of brainstorming.

- What are the needs of the local community and industry?
- Are funds available to support the selected project?
- Where are potential sources for funds?
- Which projects are of interest to the team?
- Do these projects apply and improve the team members' skills and abilities?
- For the skills that the team needs, can the team acquire them through outside assistance?

Figure 2.5 The brainstorming concept

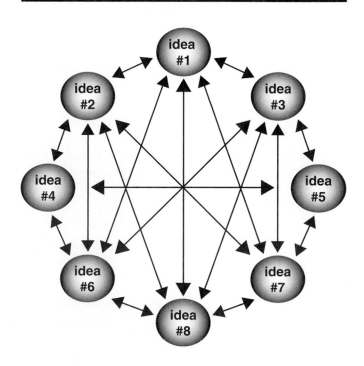

The brainstorming concept

- Approximately how long will it take to study, design, and implement the project?
- Can each project be completed within the time and funds available?

The team should try to invest a minimal amount of time in the effort of deciding on one project and see that project selection is performed as efficiently as possible.

There are, of course, other criteria depending on the projects you are considering. You can assign quantitative values (rating numbers) for these criteria to assist in the project selection process. You should try to invest a minimal amount of time in this effort and see that one specific project is selected as efficiently as possible.

Each project under consideration is evaluated in terms of the criteria. As your group evaluates its combined skills, various team members will recommend projects that can be accomplished with these skills. Your team must also identify those skills they lack. You may have to revise your list of projects and the plans associated with each one depending upon the skills of team members. Sometimes a project concept may grow beyond the funds available to implement it. Again, the plans for the project may have to be revised.

Finally, select a project that you believe is worthy of the investment in planning time, people resources, and client money.

The following are descriptions of additional approaches for solving problems. They are the bionic approach and the Delphi method.

Bionics, also known as nature analysis, uses analogies that involve animals and plants. An example of creativity that has applied the bionic approach is the study of how birds are able to fly. That information was used to design aircraft.

The **Delphi** method may be used when numeric measures are applied to potential problem *solutions.* A qualitative questionnaire is devised by a coordinator to solicit each participant's independent opinion. (The coordinator must remain neutral in questionnaire preparation, analysis, and revision.) Participants are selected whose skills are different, yet their experiences are applicable to the problem. They are asked to "weigh" or "rank" each potential solution. The coordinator then analyzes the results and distributes them again to the participants. The participants vote again on the presented alternatives, and these results are then analyzed and distributed once more. This process continues until at least 75 percent of the participants agree on the design direction to be pursued.

Brainstorming is often used to devise new (creative) solutions to selected projects. It has become a formalized approach that allows only positive contributions to devising solutions to a problem. Another approach to devising solutions is the bionic approach, sometimes known as nature analysis. These and other approaches can be evaluated using the Delphi method. Project solutions are more apt to be practical if one or more of these approaches is applied to the problem under consideration.

PROJECT EXAMPLE

Roger schedules a two-hour brainstorming session for the six team members. He lists the five potential Bedford projects. The team members shout out to Roger ideas about how the projects could be implemented. Laura writes each idea on a chalkboard under its appropriate project heading. Everyone has a good time because there are no restrictions on what can be suggested, nor are there any judgments of suggestions.

The categories and the related ideas developed by the students are listed below:

- Pave walkways

    Buy buckets of asphalt from a wholesale house and apply manually.

    Let a contractor do all the paving and spend the difference on pizza.

- Improve playground

    Smooth all rough spots with a rake and roller.

    Remove the medium- and large-size rocks; build a fence with them.

- Improve swimming pool

  Drain the pool and lengthen it with concrete.

  Construct a smaller children's wading pool near it.

  Design and install a diving board.

- Construct a children's activity center

  Design and build a wood-frame house with toilet facilities.

  Purchase a prefabricated kit and assemble it on town property.

- Replace old carousel

  Use a large version of Erector™-set materials to build a new carousel.

  Design a new carousel using parts from the old one.

Copies of this list are supplied to each team member. The team members are asked to provide their comments in writing by the next Monday.

The team develops a list of project selection criteria. The criteria include the following questions:

- What does the community need?
- Is there money available for such a project, or can the team raise the money needed?
- What are the skills and interests of the team members?
- For those skills they lack, where can the team acquire outside assistance?
- Can they complete the project before they graduate or move on to other courses?

Each of the three student groups is assigned one or two of these five potential projects. The groups are to prepare a Potential Projects document, which includes a narrative and rough estimates of the tasks, schedules, and budgets for their assigned projects. (See chapter 7 for documentation details.) They are to present their results at the beginning of week 5.

Once the team brainstorms ideas, each project must be analyzed according to a set of criteria, which includes the need for the project, the funding available, and the skills of team members.

Activity 2D: Identify Risks for Each Potential Project

Formulate a list of the problems and risks involved for each potential project. To formulate a problem is to convey your ideas via words and diagrams—you prepare a document describing your ideas and opinions. Technical problems require quantitative descriptions that can be assessed by you and others. List the areas, for each potential project, that will require later detailed analysis. Prepare a document that identifies and describes these problems to the extent that they are now known. Indicate how the risks can be evaluated during the Study Phase.

High-risk areas are those portions of a proposed solution that may be difficult to implement and may cause the project to fail. High-risk areas require further investigation. High-risk areas generally fit into one of the following four categories:

High-risk areas are those portions of a proposed solution that may be difficult to implement and may cause the project to fail. For example, new technology may need to be developed.

1. An unusual approach must be considered.
2. Advances in technology are required.
3. Personnel must develop new skills.
4. New equipment, forms, and procedures must be developed and tested.

It is essential that you study and resolve each high-risk area before you proceed with the more expensive Design and Implementation Phases. Therefore, at the beginning of the Study Phase, you must assign a high priority to the investigation of all high-risk areas.

The various projects described in the town of Bedford redevelopment plan are listed by the groups assigned to each project, along with the risks involved:

Projects	Risks
Pave walkways	Difficult material handling
Improve playground	Team lacks experience; no enthusiasm
Improve swimming pool	Chemical-handling training required
Construct children's activity center	Complicated; time-consuming
Replace old carousel	Special parts must be fabricated

Common risks for all five projects include these:

- Possible competition by local contractors
- Available funds vanishing prior to project initiation
- Lack of enthusiasm for any project by some team members

Professor Hulbert notes that Northeastern has achieved an enviable record on similar projects performed by previous classes of students. The university has been able to obtain projects because it offers work at a low cost since the students work for no wages. Thus, the school's reputation will transfer to the team's effort if the team's plans are well prepared and are convincing.

Compare the problems and associated risks of each potential project. Select one project from the list based upon your evaluation of the technical risks involved and the availability of funding. During this portion of the Conception Phase, the teams continually revise their plans as soon as significant amounts of new information or new ideas are either obtained or proposed.

Activity 2E: Select One Project and Gather More Information

Before large amounts of time and money are invested in the selection process, it is advisable to focus on one project. This is not always possible. However, one project should be selected as soon as possible, but no later than near the begining of the Study Phase.

Once the project is selected, the information-gathering process focuses on obtaining detailed information for that project.

More detailed information that directly concerns the selected project should now be gathered. Questionnaires can be developed so that interviews can be conducted with applicable personnel. Where people will be using a facility, demographic information should be obtained. Where equipment is to be designed, existing and proposed technology must be investigated. A model for simulation of all or a portion of the proposed equipment design may be necessary. (See section 9.3.) This information will be required so that practical design alternatives for implementing the project can be proposed and tested.

At the Monday meeting for week 5, the team decides to concentrate on the activity center project. It must be affordable, and there must be money available for any contracted work and for any materials that are to be purchased. A cost limit is established:

The activity center project must cost less than $20 000.

Funding must be available in advance for the purchase of materials and services that the first-year students are not able to provide. Any money remaining in excess of the overall expenditures will be donated to a Northeastern fund for student projects so that first-year students, in subsequent years, can pursue similar ventures. Professor Hulbert negotiates with the town of Bedford and the attorney for Miss Kewer's estate for the receipt of funds. The two sources inform Professor Hulbert that they will grant funds, the decision being based on the good reputation earned by the university on previous projects.

Additional data are gathered by all six team members, who now concentrate on the activity center as the team project. The work to be performed, prior to Professor Hulbert's review, includes expanding on all the gathered data (figure 2.6) and determining if other funding sources can assist in the effort, should the $20 000 not be enough.

The team evaluates the new data gathered and decides to continue with the project of constructing an activity center for local community use. The team has previously examined its skills as a group. The members now decide that the team should be able to construct a wood-frame activity center. They proceed in the following way:

Local town leaders are contacted. They offer a parcel of public land on which the activity center could be constructed. The use of private land that could be donated to the town is also explored. Several small-business contractors in the town volunteer their expertise in those areas where the team lacks the capability. Roger first discusses this information with Professor Hulbert and later with the team.

The requirements now established by the team lead it to investigate and analyze activities that can assist children in the community to develop their abilities, skills, and positive self-image. The activity center would be designed to provide activities that promote such development on the part of children who use the center.

The team next investigates ideas for children-oriented activities that could occur within an activity center structure. They interview members of the Bedford community, including school and town administrators, teachers, parents, and children. They discover that two nearby towns have structures where children can play and learn. They also discover that, while some parts of these structures are very successful, other parts of the designs needed to be changed once the structures were in use.

The team members next organize and develop a list of potential child-oriented activities for those who might use the activity center. A demographic study of the ages, gender, and interests of the children for whom the activity center is to be designed is determined by a survey. Parents whose children are in the selected age group are surveyed for their opinions and preferences.

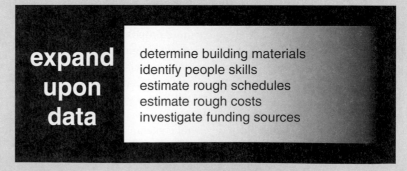

Conception Phase: additional effort for potential project

expand upon data

determine building materials
identify people skills
estimate rough schedules
estimate rough costs
investigate funding sources

Figure 2.6   Conception Phase: additional effort for potential project

The purpose of constructing the activity center becomes a combination of exercise, entertainment, and education. The team now realizes that there is more work to accomplish than there is time available. They ask Professor Hulbert for, and receive, an extra amount of time to perform these information-gathering tasks.

The consensus of the team is to offer to design and construct a wood-frame activity center. Thus, the activity center project is now definitely selected. The four phases for the activity center project are shown in figure 2.7.

A rough schedule for the activity center project is devised by Jose and Kim. One week is allotted for completing the remainder of the Conception Phase. This time includes a Conception Phase progress review to be held with Professor Hulbert.

The remainder of the week is allotted to preparing the documentation for the Conception Phase, and for preparing the estimates related to the remaining three phases. The remaining three phases are to be accomplished by the next-to-last week of August. (The last week of August is set aside for delivery of the activity center and of project documentation.) Drafts are exchanged via e-mail.

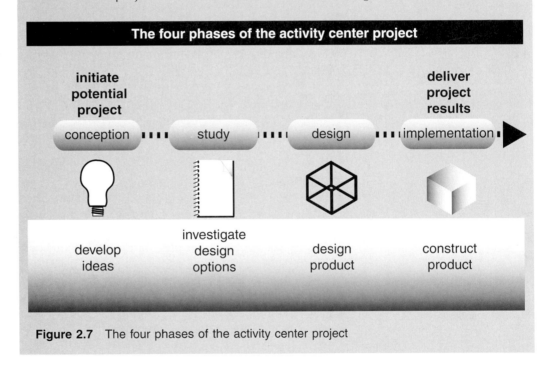

Figure 2.7   The four phases of the activity center project

Activity 2F: Search for and Document Alternative Solutions

Initiate a second brainstorming session. Prepare a list of alternative solutions for the selected project to be examined during the Study Phase. At the conclusion of this brainstorming session, identify those proposed solutions that will satisfy both your customer and your funding source. See the Project Example that follows for more detail on how this step works.

By determining the overall requirements of a project, you can apply restrictions to the results of your brainstorming session. Split your team into smaller groups so each group can pursue information that will help it establish the project requirements. Prepare an outline of a questionnaire that can be used to evaluate alternative solutions during the Study Phase. You should also establish the general content of the questionnaire at this time.

The strategy that you apply to solve a problem often depends upon your education and experiences.

* Scientists solve problems by analyzing to try to arrive at a solution.

* Designers choose solutions and try to see if one fits the problem.

Design problems cannot be stated precisely. Therefore, solutions cannot be derived directly from them. Designers have to suggest tentative solutions. The solution and problem are then explored in parallel. You need self-confidence to define and redefine a given problem. People who seek structured, well-defined problems will never appreciate the thrill of being a designer!

PROJECT EXAMPLE

A second brainstorming session is scheduled by the team members. The brainstorming results are merged, and the following list of questions related to implementing the project is assembled:

- Will the activity center require a basement?

    If there is a basement, should it be a partial or full basement?

    Should the foundation consist of concrete block or poured concrete?

- Will it be a one-story or two-story activity center?

- How many rooms should there be within the activity center?

    What uses are to be considered for each room?

    Should some activity center rooms be designed for special age or interest groups?

- Will the activity center be constructed to blend in with other local structures and the land, or should it stand out for reasons of its uniqueness?

- Are termites, snow, etc., problems in this area of the country?

Other questions can be added later as team members think of new design directions. Preliminary design requirements are established prior to extensive planning. Rough cost estimates are obtained so that the design direction can be determined during the next phase. Note that these are rough cost estimates. More detailed estimates are determined during the Study Phase (see chapter 3).

*Preliminary design require-
ments are established prior to
extensive planning.*

The establishment of requirements and design for a project is an iterative process, as will be discussed later in this book. This means the requirements are constantly being reevaluated and revised.

The general content of a design specification must now be established. Questions such as the following are asked:

- For the activity center, what locations (sites) are available?

- What are the ranges of environmental conditions, such as temperature, humidity, wind, and noise levels, that occur in this area?

- What kinds of building materials are locally available?

- What skilled labor is available locally?

- What companies and their employees might be willing to contribute labor, materials, advice, or guidance?

Some parts of the specification are initiated and written at this time. (See sections 7.1 and 8.1.) Existing local or state building codes are referenced, instead of being included in the Preliminary Specification documentation. Alice examines building codes further to ensure correct interpretation and compliance. She visits the local building, electrical, and plumbing inspectors to determine which parts of each code are stressed by the local inspectors. She also interviews contractors who are known to her father and uncle, who have worked in or near the town of Bedford. She inquires about conforming to building codes and how to determine more precisely which parts of the building codes apply to the activity center project.

Activity 3: Initiate Conception Phase Documents

Develop project plans that again include questions such as *who, what, when, where, why,* and *how:*

Who will be responsible for the work?

What is to be accomplished and by whom?

When is it to be implemented?

Where is it to be performed?

Why should it be performed?

How will the performance of the project be controlled?

The answers to these questions may have changed now that one project has been selected and more information relating to that project has been gathered.

Outlines for the documents for the Conception Phase should now be prepared. If the outlines are begun now, they can be expanded and revised to become the final documents for the Conception Phase. See activity 6 later in this chapter for a more complete discussion of the final documentation.

Activity 3A: Prepare the Project Selection Rationale

The Project Selection Rationale document explains and justifies the selected project. It also contains an outline for performing that project. In order to plan the remainder of the Conception Phase and the three remaining phases, all team members involved must assume that the work will proceed as described in this evolving document. They have examined, and are examining the following questions:

The Project Selection Rationale lists the tasks to be performed, the time required to accomplish each task, and the cost of performing each task.

1. Who investigated alternatives?

2. What led to the accepted conclusion?

3. When should one project be selected?

4. Where did we search for data?

5. Why did we select this particular project?

6. How do we plan to proceed from here?

The answers to these questions justify choosing the selected project.

The completed Project Selection Rationale document lists the tasks that were performed during the Conception Phase, explains the time and costs expended to perform them, and also explains why they were performed. (See sections 7.1 and 7.5.) See activity 6 of this chapter for an example of an outline of a Project Selection Rationale document that could be produced for the activity center project.

PROJECT EXAMPLE

Professor Hulbert advises the team to devise project plans and include revised answers to questions such as *who, what, when, where, why,* and *how.* The answers to these questions may have changed now that one project has been selected and more information relating to that project has been gathered:

Who will be responsible for the work?

What is to be accomplished and by whom?

When is it to be implemented?

Where is it to be performed?

Why should it be performed?

How will the performance of the project be controlled?

The decision to design and construct an activity center will use and develop the skills of the team. It will also provide recreation and learning opportunities for children of the community. Now the team expands the details of the *who, what, when, where, why,* and *how* questions, with the activity center project specifically in mind.

Who will be responsible for various portions of the work? Ellery, with the assistance of Alice and Kim, will be responsible for the design and construction of the activity center foundation and walls. Jose, with the assistance of Laura, will be responsible for the design layout and installation of the activity center wiring. They request assistance from the University Architecture Department on the applicability of their software.

What is to be accomplished? The team next devises a set of tasks that start with the study of potential activity center design directions (chapter 3) and end with the delivery of the completed activity center (chapter 5). The tasks are the following:

Prepare Site

Excavate Site

Erect Foundation

Cap Foundation

Erect Rough Framing (and Sheathing)

Shingle Roof

Install Windows

Install Heating

Finish Exterior Siding

Install Sheetrock™

Plaster Interior

Lay Floors

Install Interior Finish

Paint Interior Finish

Laura is assigned the responsibility for determining the risks for each of these potential tasks.

When is it to be implemented? A date is established for the official ground-breaking ceremony. The completion and delivery dates are planned to occur during week 25. Week 26 is saved for the final project documentation effort.

Where is the activity center to be constructed? Steps made of concrete will be fabricated at a supplier's site; activity center construction and assembly will occur at the Bedford Park site location. (When a circuit, device, building, or system is constructed, it is often divided into several portions, each of which is fabricated at a different location.)

Why should it be constructed? The activity center, when complete, will have provided valuable training for each team member who has participated in the project. The activity center will also add to the quality of life in Bedford by offering a place for young people of the community to exercise, play, and learn.

How will the performance of the project be controlled? Coordination and control of the entire effort will be the responsibility of Roger. All projects require that people learn to work together. All persons involved must understand their portion of the total assignment and how those portions relate to all other portions of the project. Professor Hulbert notes that Northeastern requires the responsible faculty member to review—on a weekly basis—the status, problems, and plans for resolving the problems for each project.

Activity 3B: Prepare the Preliminary Specification

The Preliminary Specification establishes the bounds of investigation for the Study Phase. Specifications traditionally fall into one of six categories:

- Residential
- Commercial
- Municipal
- Industrial
- Military
- Aerospace

Contents of typical specifications are given in section 8.1. Design specifications are considered to be complete at the end of the Study Phase.

Specifications should not be confused with **requirements.** Specifications, as described in this book, are documents that contain, among other things, a description of the work to be accomplished. Specifications include requirements, but they contain more than just requirements. Requirements are conditions that *must* be fulfilled in order to complete a task or project. Design specifications also include design preferences.

PROJECT EXAMPLE

The Preliminary Specification for the activity center is being prepared by Alice. The following parameters for the project were developed during the Conception Phase:

The activity center will be designed to the specification entitled "Specification for the Town of Bedford, Massachusetts: An Activity Center." (The final version of this document is given in appendix C.)

The activity center will be used by children from ages 4 through 10.

The activity center must cost less than $20 000.

These three sentences are the introduction to the Preliminary Specification.

Alice and Ellery develop the outline of the Preliminary Specification as follows:

General Specifications
General Requirements
Architectural Requirements
Intent of Contract Documents

Subcontractors

Project Meetings

Cleaning (of Site)

Workers' Compensation and Insurance

Project Closeout

General Site Work

Earthwork

Landscaping

Site Utilities

Concrete Formwork

Cast-in-Place Concrete

Dampproofing (Moisture Barriers)

Subsurface Drainage System

Framing and Carpentry

Roof

Window and Door Schedule

Exterior Trim and Siding

Plumbing and Heating

Electrical

Walls, Floors, and Ceilings

Locks

Landscaping, Driveway, Ramp, and Walks

Alice and Ellery are assigned to write and coordinate this document. (The Specification is to be completed during the Study Phase.)

Activity 3C: Prepare the Project Description

The Project Description document describes the Project Plan. It divides the remaining work for the Conception Phase into two parts:

1. Formal task descriptions, which include skills, schedule, cost information, and the risks involved for each task

2. The plans for the Study, Design, and Implementation Phases

See section 2.3 for requirement details in the Project Description.

Descriptions of tasks required to complete the project are now ready to be written. The tasks are arranged in a logical sequence, with time and cost estimates included.

Team members use their combined skills to identify the work to be performed and completed. (This is also known as the Work Breakdown Structure. See chapter 7.) The following procedures are a guide.

- Identify and document the work to be accomplished in the form of tasks. See appendix B.

- Decide where one task ends and another begins, based upon the ease with which it becomes obvious when the task has to be started and when the task is complete.

- Write the descriptions of these tasks so that a logical sequence exists.

- Describe, for each task, the skills required to perform that task.

- Identify and list the materials and equipment required to perform each task.

- Estimate the time required to perform each task (East and Kirby, 1990; Lewis, 1991).
- Estimate the cost to perform each task, using the various cost categories given in section 7.5.

Solicit the skills of consulting specialists, where necessary, to work with team members during the writing of task descriptions. Tasks are first listed in an outline format. The interaction of one task with all others can be expressed via a schedule that connects the start, performance, and completion of each task. Then begin initiating task *descriptions,* which are improved on a continuing basis as more information becomes available.

Miles (1972) notes that the Conception (problem-setting) Phase and the Study (problem-solving) Phase often overlap. Team members must gather much information before the actual problem to be solved can begin to be identified. The task descriptions indicate the understanding of the problem at a given moment in time.

The search for more information and problem (value) analysis continues. (See appendix A.) As the project progresses, there are fewer questions and fewer missing partial solutions. The existing problem may be divided into two or more separate problems, or changed altogether. Disciplined thinking, combined with creativity, leads to solving the remaining portions of a problem until it evolves into the *actual* problem and its solution. Thus, the Project Description is a living, changing document that evolves during the Conception Phase. It becomes the Design Description during the Study Phase. The Design Description is modified during the Study Phase as more information becomes available. The Design Description is the guide to the development of Working and Detail Drawings. The content of these drawings will be discussed during the Design Phase.

PLANNING AHEAD

For this Conception Phase, include the following for each task description within the Project Description document:

- The identity and objective of the task
- The relation of the task to the overall project
- A description of the work to be accomplished
- The skills required to accomplish each task
- The specifications that define the quality of work and materials
- A schedule for the performance of the tasks
- The costs associated with accomplishing each task

Connect the tasks to skilled individuals and organizations willing to participate in implementing the project.

Carefully review how each task interacts with all others. Identify those tasks whose completion will affect the overall costs and schedule so they can be carefully monitored.

Examine the costs for each task. How do you know what is a reasonable cost estimate? Experience is the best teacher. Therefore, retain good records of previous work and examine these records for applicability to the new project. Gather and consult catalogs containing information regarding materials and their costs. Obtain, where practical, a range of costs and schedule times for each task. Investigate alternative solutions by comparing schedule and costs. Also realize that costs control choices, and choices control costs.

Tabulate the estimated income and expenses for the project during this planning stage. This tabulation is a financial statement that summarizes the assets, liabilities, and net worth of a project, business, or individual at a given date. By preparing these tabulated data using a computer spreadsheet program, you will be able to obtain immediate information on the effect of changes on the project. (See section 6.5.) If a spreadsheet program is not available, then prepare the cost estimates manually. Only ruled paper and a calculator are necessary. However, considerable time is required, and mathematical errors are more likely to occur when the spreadsheet is prepared manually.

PROJECT EXAMPLE

The team members initiate the Project Description document for the activity center. The work for constructing the activity center must now be divided into tasks. (Tasks, their objectives, and their accompanying task descriptions are discussed in appendix B.) Laura, who enjoys writing, is given the assignment of developing a list of activity center tasks by examining the work to be accomplished. Jose and Kim work with Laura to devise an accompanying schedule. Kim, who has an understanding of bookkeeping, volunteers to obtain prices for the various materials that are to be considered. Jose agrees to investigate the scheduling software available on the University computer system. Roger assumes the responsibility for determining the design preferences of the persons who will be funding this effort.

The initial tasks, with their accompanying risks, devised by Laura, are the following:

Tasks	Risks
Prepare Site	Environmental restrictions
Excavate Site	Ledge, soil, water (such as a spring)
Erect Foundation	Temperature, weather conditions
Cap Foundation	Temperature, weather conditions
Erect Rough Framing (and Sheathing)	Quality of lumber, wind
Shingle Roof	Wind, rain, snow
Install Windows	Quality of lumber and window frames
Install Heating	Consistent availability of selected fuel
Finish Exterior Siding	Wind, temperature, humidity, rain
Install Sheetrock™	Humidity, temperature
Plaster Interior	Humidity, temperature
Lay Floors	Condition of rough flooring
Install Interior Finish	Temperature, humidity
Paint Interior Finish	Temperature, humidity

Laura realizes that this list of tasks is likely to be revised during the Study and Design Phases.

Activity 3D: List Tasks, Schedule, and Budget Constraints

All projects are finite in time and costs. Identify those parts of a planned project that are most likely to deviate from your plan or will require further study. Identify especially any areas that will require detailed analysis during the Study Phase. These areas, along with the involved time and costs, are the constraints on this project.

Approximate schedules need to be devised based upon the list of tasks developed. See chapter 7 for information on this topic.

Approximate costs and schedules are developed based on the initial list of tasks needed to implement the project.

The expression **contingency fee** is used in cost estimating to cover unforeseen expenses. It describes the amount of money that is set aside to cover the cost of complications that are suspected or likely to arise, but have not yet been isolated and resolved. For the Conception Phase, these categories are the project uncertainties. The greater the expectation that all design complications have not yet been identified and resolved, the greater the contingency amount.

It is customary to add a percentage to the initial cost estimates to include items whose complexity is not yet understood. The greater the risk involved, the greater the contingency percent. This helps ensure that the final estimates, which are determined during the Implementation Phase, are likely to be within the original estimates determined during the Conception Phase. The actual percentage chosen depends upon the experiences of the estimators and the degree of uncertainty of the data used for estimating.

PROJECT EXAMPLE

The tasks to implement the project, with their associated risks, have been previously listed by Laura. Jose and Kim work with Laura to devise a rough schedule, using her initial list of tasks for the Implementation Phase. The time indicated to accomplish each task is noted to the right of each task in figure 2.8.

Note that the schedule requires approximately six months to accomplish the Implementation Phase. This is a constraint imposed by the school year.

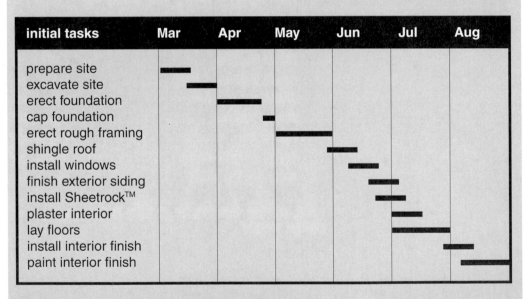

Sample of a rough schedule for the Implementation Phase

initial tasks	Mar	Apr	May	Jun	Jul	Aug

prepare site
excavate site
erect foundation
cap foundation
erect rough framing
shingle roof
install windows
finish exterior siding
install Sheetrock™
plaster interior
lay floors
install interior finish
paint interior finish

Figure 2.8   Sample of a rough schedule for the Implementation Phase

Kim works with the other members of the team to obtain costs for the various materials to be considered during the Study Phase. Jose and Laura work to convert the materials costs into a total budget, which includes labor and other categories, as discussed in section 6.6. For each entry in the rough cost estimates given in figure 2.9, the students include appropriate contingency amounts as advised by Professor Hulbert. (See chapter 7.)

Note that the total money available is $20 000; the extra $800 (4%) is for contingencies. This is a constraint imposed by the contributing funding sources. It is assumed that there will be no significant costs associated with the Study and Design Phases. The team members and Professor Hulbert are contributing their labor.

↓ ITEM / MONTH →	MAR	APR	MAY	JUN	JUL	AUG	TOTALS
Initial Tasks							
prepare site	600						0
excavate site	1400						0
erect foundation		1700					1700
cap foundation		400					400
erect rough framing			4800				4800
single roof				1000			1000
install windows				1800			1800
finish exterior siding				400	400		800
install Sheetrock™				300	500		800
plaster interior					600		600
lay floors					2200		2200
install interior finish					600	1200	1800
paint interior finish						1300	1300
totals	2000	2100	4800	3500	4300	2500	19200

When using an electronic spreadsheet, it is possible to insert a check cell (I17) to verify that the entries were totalled correctly, because the horizontal row and vertical column are the same. Changes in any of the values will automatically change the two totals.

	A	B	C	D	E	F	G	H	I
1	↓ ITEM / MONTH →	MAR	APR	MAY	JUN	JUL	AUG	TOTALS	
2	Initial Tasks								
3	prepare site	600						0	
4	excavate site	1400						0	
5	erect foundation		1700					1700	
6	cap foundation		400					400	
7	erect rough framing			4800				4800	
8	single roof				1000			1000	
9	install windows				1800			1800	
10	finish exterior siding				400	400		800	
11	install Sheetrock™				300	500		800	
12	plaster interior					600		600	
13	lay floors					2200		2200	
14	install interior finish					600	1200	1800	
15	paint interior finish						1300	1300	
16	**totals**	2000	2100	4800	3500	4300	2500	19200	across:
17								(down)	19200

Sample of a rough cost estimate for the Implementation Phase

initial tasks	Mar	Apr	May	Jun	Jul	Aug	totals
prepare site	600						600
excavate site	1400						1400
erect foundation		1700					1700
cap foundation		400					400
erect rough framing			4800				4800
shingle roof				1000			1000
install windows				1800			1800
finish exterior siding				400	400		800
install Sheetrock™				300	500		800
plaster interior					600		600
lay floors					2200		2200
install interior finish					600	1200	1800
paint interior finish						1300	1300
totals	2000	2100	4800	3500	4300	2500	19 200

Figure 2.9 Sample of a rough cost estimate for the Implementation Phase

Activity 4: Expand Team and Identify Supporting Organizations

Toward the end of the Conception Phase, the skills required to perform the chosen project are reevaluated, as are the skills of existing team members. New team members may be added, and sources of outside help are identified.

Identify additional persons who want to be members of the team for the remainder of the project. Establish a formal organization of all team members. Decide upon a time and place for these new persons to get together and meet the existing team members. This initial meeting is often referred to as a project "kickoff meeting."

Verify your choice of team leader. For some projects, the team leader may be appointed by the organization involved in overseeing the project. The team leader should have both technical and managerial skills and should be a decisive person. For groups of individuals whose focus is on learning to work together, it may be desirable to periodically change the team leaders. This provides good training for those who want to test their management and organizing skills. It also allows the team members to become acquainted with different management styles.

Designate the team leader as **manager** if the team size exceeds nine persons. (This principle derives from management texts and is referred to as the "rule of ten.") At this point, skills and abilities needed to complete the project are reexamined, and additional ones needed are identified. Once again, the manager may need to invite additional persons to become team members. New members are often recommended by existing team members. Projects conceived by college teams may attract retired individuals, who may contribute their skills at a very low cost or at no cost. There also may be other options that could be considered.

Identify other persons or organizations who have the additional skills and abilities required to perform the planned project. These persons or organizations may become involved via a contract, as discussed in section 8.3. Someone may discover that the required additional skills are available from a local company that has access to funds available for the planned project.

PROJECT EXAMPLE

It is time to examine the composition of the project team. The initial six team members examine their skills and abilities.

Roger: Enjoys the leadership role and is accepted by his fellow students; he wants to locate a spreadsheet program that will assist him in comparing the estimated time and costs with (later) actual expenditures.

Alice: Wants to work on the actual design; her father has volunteered the time of a civil engineer who can advise in site layout and heavy-equipment requirements. She is comfortable with converting the Conception Phase ideas into designs during the Study Phase.

Laura: Is creative and enjoys organizing information; she realizes that she has no idea as to how much time will be required to perform the work of each task. She locates a third-year student in accounting to assist with the next phases.

Jose: Likes to work with his hands; he needs assistance to determine sources of information regarding foundation forms and how they are used.

Kim: Is interested in chemistry; she needs information regarding lumber and concrete. There is no first-year student with that information. She contacts a fourth-year student who desires to specialize in these areas during his graduate-school studies.

Ellery: Has drafting experience; he would like to advance his skills by utilizing a computer CAD program for preparing the Study and Design Phase drawings. He seeks advice from Jose.

The team members discover that, to continue, they must add new members to the team for the Study Phase. Where there are no skills or abilities available within the university, they must develop contacts with potential contractors.

Two more first-year students are added to the team in preparation for the Study Phase: Manuel Cardoza, a computer technology major with software experience; and Joseph O'Neil, formerly a carpentry student at a vocational high school.

At the progress review meeting, held with Professor Hulbert, these two new members meet the other team members, are informed of the status of the Conception Phase, and receive their assignments for the Study Phase. Professor Hulbert addresses the issues involved with expanding the six-member team to an eight-member team. He also notes that the coordination required among

> members for teams containing more than three persons can be significant. The individuals in a larger team have fewer individual responsibilities. However, the increase in team size requires an increased coordination effort.

Activity 5: Complete the Planning of the Next Phases of Work

Plans should be realistic with regard to timely performance and reasonable cost. Meeting these criteria ensures that the actual project will be performed successfully within the scheduled time and financial constraints.

Good planning requires (1) that there is sufficient information available for planning purposes and (2) that the person or persons planning have sufficient experience to use the information wisely. Plans should be realistic with regard to timely performance and reasonable cost. Meeting these criteria ensures that the actual project will be performed successfully within the scheduled time and financial constraints. Information useful to planning, such as actual schedules and costs from similar projects, is often gathered during the Conception Phase. It is then available for use during the Study, Design, and Implementation Phases.

Determine the approximate amount of work necessary to perform the Study, Design, and Implementation Phases. Prepare schedules and estimate costs for performing the remaining activity (final documentation) of the Conception Phase, review all of the activities for the Study, Design, and Implementation Phases, and prepare documentation. The documentation at this point should be restricted to rough estimates until the Study Phase is in progress.

The documents for the Design and Implementation Phases are considered to be approximate in content. They are often labeled "Draft," because further information is necessary to expand and complete plans. The time and cost to do more in-depth evaluations, and the effect of these evaluations on the project, will be examined in more detail during the Study Phase.

PLANNING AHEAD

The managers and reviewers within the organization will want to know that the project will cost no less than, and no more than, specific dollar amounts. Establish the boundaries of low and high project costs before beginning the Study Phase. This process is referred to as **bracketing** the project costs.

- The low-end costs can often be established by examining the data from previously completed projects whose tasks are comparable; unusual development effort is not required.

- The high-end costs can be roughly estimated by interviewing those persons who will be responsible for performing those new tasks whose content requires research and development effort. It then becomes the responsibility of the manager to decide upon a final price for the project. Managers should include contingency funds (see activity 3D) where they suspect the research and development personnel have not provided realistic estimates.

It then becomes the responsibility of the manager to decide upon a final price for the project. Why? Because most funding agencies respond more favorably when only one amount of money is requested. For a funded project, enough money must be earned in excess of expenditures to fund the first stages of future projects that may be identified and proposed later.

Often, funding for only the Study Phase will be authorized. The reason for withholding funds for later phases is that persons responsible for allotting funds want to be assured that the cost of the later phases will not exceed the available money.

Identify any areas that may require detailed analysis during the Study Phase. The time and cost to evaluate these areas, and the impact the results will have upon the project, must be identified now. As an example, human-factor requirements should be included. (See section 9.2.) Revise all plans as soon as significant amounts of new information are obtained or new ideas are conceived and developed.

At some point, the chosen leader must decide when the Conception Phase will end. It is possible that new information may be gathered during the Study Phase that may require briefly reviewing the results of the Conception Phase.

PROJECT EXAMPLE

Planning the activity center project requires the development of a flowchart showing the sequence of the tasks, as shown in figure 2.10.

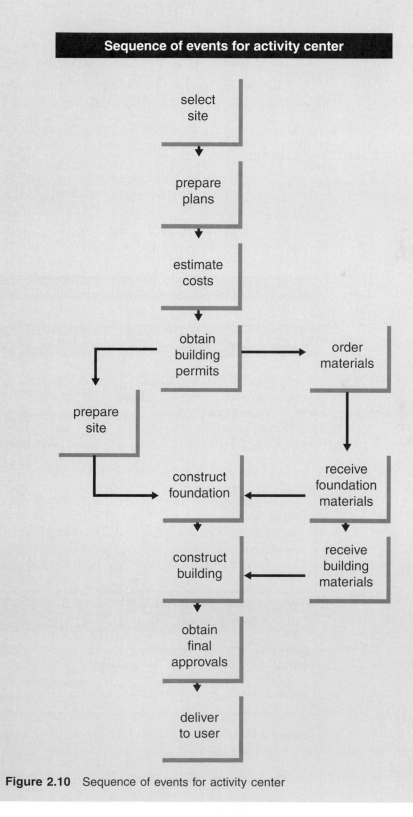

Figure 2.10   Sequence of events for activity center

The purpose of planning is to ensure that the project, including the drawings and specifications, meets the following criteria:

- The proposed structure is in compliance with the zoning regulations. (Zoning approval is required before a construction permit can be issued.)
- The site work, clearing, and rubble disposal plans are acceptable to authorities.
- The activity center location meets local building codes and other requirements.
- A realistic base is used for determining costs.
- On-schedule completion of the job can be accomplished.
- The project can be completed with the available funds.
- The work is acceptable to the funding sources.

These factors are interrelated. Each day the world becomes more complicated as there are more restrictive regulations and guidelines to consider.

The time between mid-November and the end of December is allotted to the Study Phase. The months of January and February are assigned to the Design Phase. The sequence of events for the project is devised and reviewed. The overall schedule for the project is shown in figure 2.11.

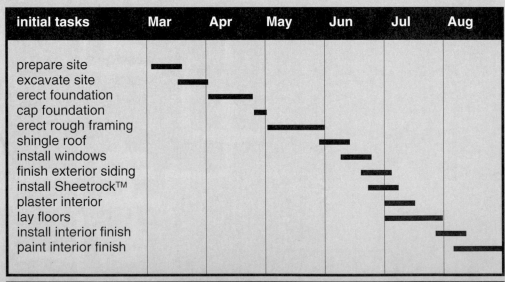
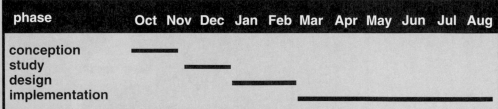

Figure 2.11   Overall schedule of project phases

Activity 6: Convert Document Drafts into Final Documents

Documents for the Conception Phase typically include the following:

- *One that explains why the project was selected*
- *One that contains all the specifications for the project*
- *A general document that describes the project and that contains preliminary proposed costs and schedules*

A phase is not finished until the documentation (paperwork) is complete. Here are the documents to be completed during the Conception Phase:

1. Project Selection Rationale
2. Preliminary Specification
3. Project Description (with an artist's concept)

Simple projects may merge these documents. More complicated projects may require additional documents. The content of each of the above documents is described in section 2.3. These documents can be modified to become portions of the Study Phase documents; see section 3.3.

CONCEPTION

PROJECT EXAMPLE

As noted above, documentation of activities should be prepared as each of these activities occur. Documentation should *never* wait until all the activities of the Conception Phase are completed.

All involved team members and Professor Hulbert agree that the document for Project Selection Rationale for the activity center project needs to consist of the following:

1. A description of the potential projects explored
   1.1 Pave existing walkways
   1.2 Improve playground
   1.3 Improve swimming pool
   1.4 Construct children's activity center
   1.5 Replace old carousel
2. Attributes of each potential project
3. Approximate tasks, schedules, and budgets for each project explored
   3.1 Pave walkways: $4 500 requiring nine weeks
   3.2 Improve playground: $4 200 requiring fourteen weeks
   3.3 Improve swimming pool: $7 500 requiring ten weeks
   3.4 Construct children's activity center: $20 000 requiring twenty-six weeks
   3.5 Replace old carousel: $28 500 requiring forty-two weeks
4. Justification of the project selected
   4.1 The need by the town of Bedford for the project
        Why is the activity center needed?
        What activities will occur in the activity center?
        How will these activities meet the needs?
   4.2 The schedule and money available
   4.3 The complexity of the project
   4.4 The skills of the team
   4.5 The type of technical and administrative assistance required

Written text, with the appropriate charts and other figures, accompanies each of the above descriptions. Laura is assigned to coordinate the writing of this document. She requests the e-mail addresses from each team member.

For discussion of the Preliminary Specification, see section 2.3.

Figure 2.12   Artist's concept of student activity center

The Project Description document for the activity center contains the following:

- Justification for the project (from the Project Selection Rationale document)
- Semitechnical description of the range of Study Phase options, including an artist's concept of the activity center. (See figure 2.12.)
- Estimated schedule for the Study, Design, and Implementation Phases
- Estimated costs for the Study, Design, and Implementation Phases
- The sources of funding to achieve the $20 000 needed to implement the project
- Directions for the Study Phase

    Ideas to be investigated

    Building-size options

    Types of siding

    Types of roofing

    Landscaping

    Questions to be answered

        What locations in Bedford Park are available?

        What are the ranges of environmental conditions, such as temperature, humidity, wind, rain, snow, and noise levels, of this area?

        What types of building materials are available locally?

        What skilled labor is available locally from companies and employees who might be willing to contribute labor and materials—or perhaps just advice and guidance?

What types of lumber and concrete should be considered?

What foundation construction techniques are applicable?

What are the Bedford water and sewage connection requirements?

Where is electrical power available during construction?

- Staffing skills required for the Study, Design, and Implementation Phases

  internal: Manuel and Joseph

  external: materials vendors and construction subcontractors

- Resources needed for the next phases:

  Facilities

  Equipment

  Computation support

Ellery is assigned to prepare artist's concepts of the activity center. These are used to explore the design options during the Study Phase. Roger agrees to coordinate the writing of this document and to write the justification paragraph.

It is the first Monday in November. Professor Hulbert informs the team that the Conception Phase must end in eleven days.

SECTION 2.3   CONCEPTION PHASE DOCUMENTS

The Conception Phase begins when an idea is conceived or a need is identified. There may be more than one idea or need and several solutions being considered. Eventually, the project team selects one idea or need and begins to develop a project. The flow from the idea or need to the Conception Phase documents is shown in figure 2.13.

Team members compare their concepts of how the project should progress. Papers are written, circulated, and criticized. Meetings are held to compare the various concepts; meeting reports (see section 8.3) are written. These papers and meeting reports eventually lead to the three Conception Phase documents (figure 2.13).

At the end of the Conception Phase, the completed documents are the following:

1. The Project Selection Rationale: an explanatory document

2. The Preliminary Specification: the beginning of the specification for the project

3. The Project Description with Artist's Concept: the first drawings and plans

The content of each of these documents is described below. (More complicated projects and programs may require additional documents.)

The **Project Selection Rationale** consists of the following:

1. An analysis of the market, where applicable

2. A description of the potential projects explored

3. The attributes of each potential project

4. Approximate tasks, schedules, and budgets for each project explored

5. Justification for the project selected

This document typically consists of less than fifty pages for a modest-size project. For a large program, this document may consist of several volumes of information.

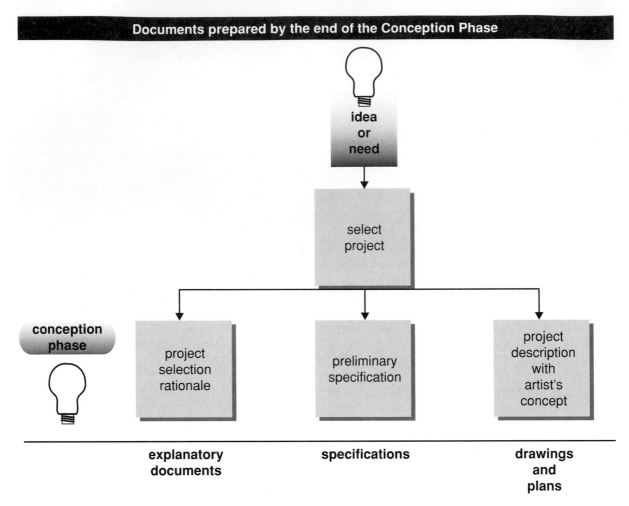

Figure 2.13   Documents prepared by the end of the Conception Phase

The Conception Phase documents contain preliminary information about a project:

- *The **Project Selection Rationale** explains why the project was chosen and contains a preliminary list of tasks, schedules, and budgets.*
- *The **Preliminary Specification** document establishes the bounds of investigation for the Study Phase.*
- *The **Project Description** contains the first drawings or plans.*

The **Preliminary Specification** document establishes the bounds of investigation for the Study Phase. Specifications traditionally fall into one or more of these four categories:

1. Residential or Commercial
2. Municipal or Industrial
3. Military
4. Aerospace

Specifications are documents that contain, among other things, a description of the work to be accomplished. There are standard contents for specifications written for each of these categories. Specifications are described in section 8.1.

Specifications should not be confused with requirements. Specifications contain requirements, as indicated in chapter 8. Requirements are conditions that must be fulfilled in order to complete a task, project, or program. However, specifications may also contain preferences for the way the project is to be implemented.

The **Project Description with Artist's Concept** consists of the following:

1. Justification for the project from the Project Selection Rationale document
2. A semitechnical description of the project that nontechnical financial persons can comprehend

3. An estimated schedule per phase

4. Estimated costs per phase

5. Potential sources of funding

6. Directions for the Study Phase

    ideas to be investigated

    problems to be explored

7. Staffing skills required for the next phases

    internal: persons either available or to be hired

    external: vendors and subcontractors

8. Resources needed for the next phases

    facilities

    equipment

    computer support

The artist's concept of the selected project's physical appearance should be included where applicable. For software projects, a sample of the expected hard-copy output is often supplied. For physically large projects or programs, a scale model of the artist's concept is often constructed, or a computer model may be devised.

When preparing three separate documents, it is important that the contents overlap. For small projects, only one combined document is necessary.

SECTION 2.4

Sometimes the end of the Conception Phase occurs when the documentation for the phase is completed; the manager must decide when the phase is complete, or a schedule may signal its completion.

DEFINING PHASE COMPLETION

The moment when the Conception Phase can be considered complete may not be obvious to those directly involved. Sometimes its end is defined by the completion of the required documents (see previous activity 6). Other times, the team may continue to examine potential projects or continue to evaluate the cost of the chosen project.

The team could now be split into two groups. One group, primarily consisting of creators, would continue to explore alternatives. (These efforts are known as *feasibility studies*.) The second group, primarily consisting of performers, would begin to study one or two selected projects and their potential solutions. This approach, however, can be very dangerous because it can lead to conflicts between the two teams. Also, any new ideas that the first group of creators presents may cause the performers to realize that they have been performing tasks that may no longer be relevant to the project.

The manager must:

*   decide when and how the Conception Phase will be terminated,

*   insist that all participants proceed on mutually acceptable paths, and

*   be certain that all plans are documented in the appropriate reports described in activity 6.

The Conception Phase schedule should indicate specifically when each output document is to be ready in draft form, when the review cycle will occur, and when these documents are ready to be published.

CHAPTER OBJECTIVES SUMMARY

Now that you have finished this chapter, you should be able to:

1. Describe the purpose and goal of the Conception Phase of a project.

2. Explain the tasks involved in the Conception Phase of a project and the six activities or steps through which these tasks are performed.

3. Describe the process of selecting a project, including brainstorming and the criteria for evaluating projects.

4. Describe the process of organizing a team, utilizing skills of team members, and expanding the team when needed.

5. Explain the nature of risks involved in projects and how to search for alternative solutions.

6. Explain the process of listing tasks, schedules, and budget constraints for a project.

7. Describe the documents prepared during the Conception Phase.

EXERCISES

2.1   Revise the Purpose and Goal statement given in section 2.1 to match each of your (multidisciplinary) class project options.

2.2   Examine the activities listed and described in section 2.2 carefully; adapt them to each of your options for a class project.

2.3   Select a team leader by secret ballot. Assist that chosen leader in formulating a team for developing the class project options.

2.4   Devise a matrix for comparing the skills and interests versus the specialty of each potential team member. Apply weights, from 1 to 5, for each entry in the matrix, where 5 is the most important and 1 is the least important. Use these weights to assign team members to portions of the Conception Phase.

2.5   Develop a project concept statement that requires minimal support from outside the team. Expand upon the ideas offered and gather additional information associated with these ideas.

2.6   Identify projects that are related to the initial list of class projects under consideration. Brainstorm to develop plans that answer the questions: *who, what, when, where, why,* and *how.*

2.7   Identify those individuals who will be the most capable of judging the details of the initial list of class projects.

2.8   Visit those individuals or organizations that might participate in the funding of each of the projects on the initial list of class projects. Prepare a written report that describes each visit.

2.9   Initiate the development of a list that can be used to establish the actual needs, desires, and luxuries for each of the customers who indicate they have a need to be fulfilled. For each entry, indicate the risks involved in meeting these needs.

2.10   Match the list of potential funding sources with your initial list of class projects. Prepare a simple presentation that describes each project from your initial list of class projects. Visit the potential funding sources and deliver your presentation.

2.11   Convene a meeting of all team members with your faculty advisor. Discuss all of the project selection criteria listed in activity 2C. Use these criteria to assist in selecting a project for each team.

2.12   Formulate the problems for each project and identify the associated risks. Attempt to answer these questions:

Are we certain that we have identified the problems to be solved?

Are these the actual problems or the perceived problems?

How can we distinguish between the actual and the perceived problems?

2.13   Select one project from that list based on your evaluation of the technical risks and the availability of funding. Search for and propose alternative solutions to the selected project.

2.14   Identify high-risk areas for the selected project. Indicate the effort necessary to analyze those areas during the Study Phase.

2.15   Determine the work to be accomplished that will lead to the proposed solution. Develop approximate tasks, schedules, and budgets for the Conception Phase work. (See appendix B.) Indicate any constraints that must be imposed upon the planned effort.

2.16   Enhance your project team to fulfill the work to be performed during the Study Phase. Identify any supporting organizations needed to perform your project. Contact those organizations and verify their willingness to participate.

2.17   Prepare the approximate tasks, schedules, and budgets for the Study, Design, and Implementation Phases. Establish low and high project cost estimates. Identify those areas that will require detailed analysis during the Study Phase.

2.18   Document the proposed solutions for the selected project and the reasons for their choices. Prepare the three documents described in activity 6.

2.19   Identify the responsibilities for each member of the project team and for any supporting organizations.

2.20   Identify the actual sources that will fund your selected project. Prepare and deliver a presentation to them. Obtain written verification that they will fund your project, including a list of their criteria for project-funding acceptability.

2.21   Review all of the Conception Phase effort. Complete all documents referred to in the exercises above. Terminate the Conception Phase.

2.22   Describe how you would identify the optimum time to terminate the Conception Phase.

CONCEPTION

chapter *3*

1. What are the purpose and goal of the Study Phase of a project?

2. What tasks are involved in the Study Phase of a project, and what activities or steps are performed to complete these tasks?

3. What is the process of reviewing the Conception Phase and planning for the Study Phase?

4. How can team members' skills be utilized to complete the Study Phase, and what can be done to manage team members so they keep to the overall goals of the project?

5. What are the ground rules for a project, and how they are established?

6. How are various solutions for implementing a project investigated and studied, and how are the associated risks and trade-offs evaluated?

7. How can the selection among design alternatives be performed as a group process?

8. What relationships are there with funding sources and potential vendors during the Study Phase?

9. What documents are prepared during the Study Phase, and how do they evolve from the Conception Phase documents?

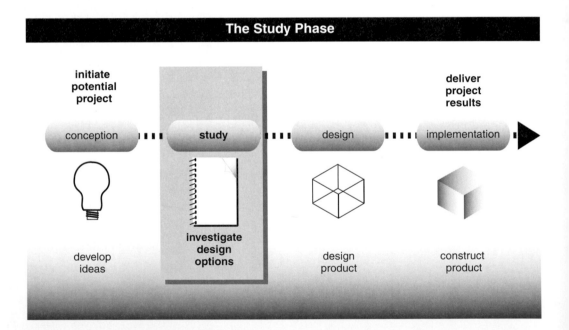

Crafty men [despise] studies; simple men admire them; and wise men use them.

Sir Francis Bacon (1561–1626)

This chapter evaluates the possible directions that the project may pursue. It is less costly to evaluate these directions at this time and select one approach to the following phases.

SECTION 3.1   STUDY PHASE PURPOSE AND GOAL

During the Study Phase, the ways to implement the project chosen in the Conception Phase are studied and developed.

The **purpose** of the Study Phase is to develop and evaluate ways to implement the project selected during the Conception Phase. The **goal** of the Study Phase is to produce a study that effectively directs the final design of a chosen project, which is developed in the Design Phase. The probability of successfully completing a project is lowest, and hence the risk and uncertainty are highest, at the start of a project. The probability of successful project completion should improve as the Study Phase continues.

At the end of the Study Phase, you should know how to accomplish the following:

*   Review the Conception Phase results and organize the Study Phase work.
*   Establish alternative solution requirements and accompanying constraints.
*   Devise criteria for judging the Study Phase outcomes that are compatible with the expectations of the client and funding sources.
*   Develop solution alternatives that satisfy the devised criteria.
*   Identify and evaluate risks and trade-offs for each proposed alternative.
*   Contact and involve vendors and subcontractors that will support the team.
*   Prepare documentation describing proposed design alternatives.
*   Refine previously devised task descriptions, schedules, and cost estimates for the Design and Implementation Phases.
*   Obtain approval, and funding, for the Design Phase.

During the Study Phase, you must continually revise all plans related to your selected project as you develop new solutions.

Note: The Study and Design phases may overlap. See section 3.4.

SECTION 3.2   STUDY PHASE ACTIVITIES

The activities for the Study Phase consist of the following:

Activity 1: Organize the Study Phase Work

    1A: Review the Conception Phase effort.

    1B: Develop the Study Phase details.

    1C: Organize teams into small groups.

    1D: Assign each group a set of tasks to investigate.

Activity 2: Establish Ground Rules for Study and Design Phases

Activity 3: Study Solutions to Be Considered for Design

    3A: Study potential solutions.

    3B: Determine functions to be included in the solutions.

    3C: Determine risks and trade-offs.

Activity 4: Gather and Evaluate Information

    4A: Contact potential vendors for their input.

    4B: Evaluate design alternatives and refine estimates.

Activity 5: Select and Plan One Solution

    5A: Select a solution.

    5B: Plan the Design and Implementation Phases.

STUDY

Activity 6: Document the Selected Solution

   6A: Refine the task, schedule, and budget documents.

   6B: Document the proposed solution and rationale.

Activity 7: Verify and Obtain Funding

   7A: Verify actual availability of funding.

   7B: Obtain Design Phase funding.

As is true for the Conception Phase, the purpose and goal you define for the Study Phase determine the activities that you are to accomplish. You may change the sequence of the activities described in this chapter from one project to another. However, you must accomplish all of the activities before the Study Phase is complete.

For small, well-defined projects that require a minimum of conception and study, the Conception Phase may be merged with the Study Phase. However, all the activities must be examined and considered. For those projects that require a minimum of study and design, the Study Phase may be merged with the Design Phase. Again, all activities must be considered.

Activity 1: Organize the Study Phase Work

During this activity, you will focus upon reviewing the Conception Phase documentation, developing the details for the Study Phase, organizing the team members into small groups, and assigning each group a set of tasks.

Activity 1A: Review the Conception Phase Effort

Team members need to review documents from the Conception Phase so that everyone understands the overall nature of the project and the proposed solutions.

Read and review the content of the Project Selection Rationale, Preliminary Specification, and Project Description documents. Some statements in them may seem new to you because time has elapsed since they were written. Also, new members may have been added to the team for this phase. At this time, the team needs to evaluate the results of any feasibility studies (alternative approaches) as well. Some team members may be stimulated to suggest new ideas or paths for the Study Phase, many of which can be very useful to the project. Review the tasks, schedules, and budgets prior to the start of the Study Phase work. Prepare and issue a memorandum, if necessary, that indicates your revisions to the original Study Phase plans.

You must now review these questions:

Who will be responsible for each portion of the work?

What is to be accomplished and by whom?

When is it to be implemented?

Where is it to be performed?

Why should it be performed?

How will the performance of the project be controlled?

You must determine if the answers to these questions (figure 3.1) have changed from the Conception Phase.

Request a meeting and review project status on a periodic basis. This is often referred to as a progress or status review meeting. If the team is a large one, the group leaders may be the only ones who attend this review. At such a progress or status meeting, you are all expected to express your opinions regarding the direction being followed. Project redirection should also be considered.

The word consensus implies that all the persons involved in the project agree on a common goal. It is the best approach for project decision making because all persons involved are more willing to work towards that common goal.

Work towards consensus. The word *consensus* implies that all the persons involved in the project agree on a common goal. It is the best approach for project decision making because all persons involved are more willing to work towards that common goal. However, factors over which designers have no control may require the managers to select a non-consensus option.

Figure 3.1   The continuing key questions for a team

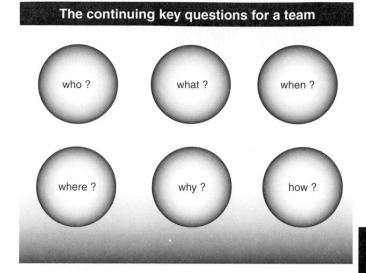

The continuing key questions for a team

who ?   what ?   when ?

where ?   why ?   how ?

STUDY

PROJECT EXAMPLE

It is now Friday afternoon in mid-November. Manuel and Joseph have been added to the team since the Conception Phase. The team members meet after classes to assess their school assignments, note their mutual availability, and plan the beginning of the Study Phase.

Roger Stein provides copies of the three Conception Phase documents to each Study Phase team member. Their weekend assignment is to read them and list any changes that they feel should be considered. These lists, and any other comments, are to be delivered to Laura, via e-mail or compatible computer disks and programs, by 8 A.M. Monday morning. Laura has no morning classes and has agreed to incorporate all lists and comments into one memorandum.

Roger has a separate list of Study Phase tasks; he provides a copy to each team member. Each team member is asked to rate these Study Phase tasks from 1 (least interesting) to 5 (most interesting), according to personal preferences. Roger collects these lists.

Activity 1B: Develop the Study Phase Details

Review the planning information that was gathered during the Conception Phase. Determine if the Study Phase tasks (appendix B), schedules, and budgets should be revised. Constrain the schedules to a time span, and budgets to an amount, that is equal to or less than was estimated during the Conception Phase.

Prepare the tasks in an outline format. The interaction of each task with all others can be expressed via a schedule. The schedule connects together the start, performance, and completion of each task. (See chapters 6 and 7.) Task descriptions are begun and improved on a continuing basis as more information becomes available.

Continually examine the cost of the overall project, particularly those of the Design and Implementation Phases. As the Study Phase progresses, the complexity of your project usually becomes more apparent. With a growth in project complexity, costs will grow proportionally. If the money required to complete the project increases significantly, then there is a greater chance that the project will be terminated by the funding source.

PLANNING AHEAD

Visit the personnel at the funding source on a periodic basis. As the project progresses, sometimes the people at your funding source will have decided to invest the money set aside for your project in a more advantageous project, may have changed their priorities, or perhaps the money they allotted is not available to them as quickly as they had expected. Thus, even if

Plans should be devised so that they are versatile and so that one or more tasks may be eliminated without a serious effect on the project. This will allow for completion of a project if some expected funding is lost.

your project team has stayed within their previously allotted funds, those funds may no longer be available.

Devise your plans for the later phases so that there is some versatility to those plans. These are often referred to as **design options** or **planning with performance measures** (Nadler, 1970, p. 768). The tasks are arranged in such a way that one or more of them may be eliminated without a serious effect on the entire project. That is a reason for continuing to plan your design functions so that they are arranged according to the apparent needs, desires, and luxuries as perceived by the people within your funding source. Avoid the temptation to prejudge the changing priorities of your client but continually monitor those changes.

The versatility of your plans for the later phases should not be openly noted in any presentation that precedes later-phase funding. Otherwise, the personnel at the funding source may decide to retain some of the funds desired for your project and let you work with a less versatile design.

Continue to compare your Study Phase project expenditures with your total accomplishments and overall goals of the projects. The project manager is held responsible for the effects of accumulated costs and projected cost estimates. Time also represents money. How much time should be allotted for project design and implementation? The longer the time a project requires—from conception to delivery—the greater are its costs. As noted in chapter 2, the written documentation must be clear and concise. It is the only material usually available to the project evaluators at the funding source.

PROJECT EXAMPLE

The memorandum prepared by Laura is distributed at the Monday afternoon meeting. After three hours of intense discussion, its content is merged into one flowing document to be printed and given to all team members the next day. Also added to the memorandum are the expanded details generated during the Monday afternoon meeting regarding the team's new understanding of the Study Phase work. Jose agrees to merge the newly developed details into the Study Phase task descriptions, schedules, and budgets.

Activity 1C: Organize Teams into Small Groups

One of the important steps in the Study Phase is to organize the work to be done during the phase and to assign team members to the various tasks.

Examine the work to be accomplished during the Study Phase, as modified during the above activities 1A and 1B. Examine further the skills, compatibility, and availability of your team members. Organize your team into small groups in preparation for activity 1D. Choose a leader for each group.

It is the project manager's responsibility to assign people to the activities to be performed during the Study Phase. At this time, the manager may also decide to add more people with a wider variety of skills to the Study Phase team. Recall from chapter 2 that the specific types of skills required for the Study Phase are more performers and fewer creators and that the coordination required among members for teams containing more than three persons can be significant (figure 3.2).

You need to consider two groups when planning the Design and Implementation Phases:

- One group for project needs for these two later phases
- One group for your client who will be using the final product

These two groups of people, working together, will contribute to a more orderly product transfer at the end of the project.

The plans for your project groups may lead to the hiring of new members or reassignment of some of your existing team members. It may lead you to search further for vendors who are more adaptable and less costly than are other members of your team. Consider employing some

Figure 3.2 Types of skills needed for each phase

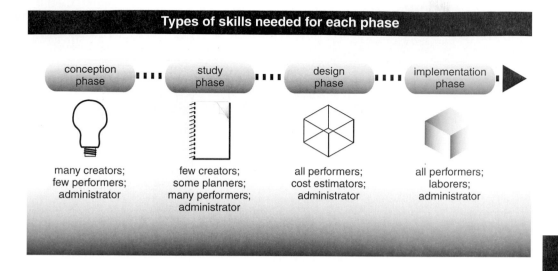

Types of skills needed for each phase

conception phase — study phase — design phase — implementation phase

many creators; few performers; administrator

few creators; some planners; many performers; administrator

all performers; cost estimators; administrator

all performers; laborers; administrator

STUDY

of the people who will eventually use your product as members of your Design Phase team or at least include some of them on your periodic review panel. Thus, they will gradually accept responsibility for ownership of the project.

If there are too many persons involved during the Study Phase who are primarily creators, they may continue to conceive new ideas. That may inhibit the progress of the Study Phase. It is important to maintain the balance between creators and planners for a given project (figure 3.2).

Progress begins by getting a clear view of the obstacles. Lenkurt Electric Co., Demodulator, Vol. 19, No. 4

Organize your project team into small groups. Ask the following questions to assist in this process:

- What are the necessary skills for performing the Study Phase?
- Which of the team members' skills apply to completion of the Study Phase?
- What local applicable skilled labor could assist you?
- Who would supervise the work?

In all probability, the composition of your groups will change as the Study Phase progresses.

PROJECT EXAMPLE

During the weekend, Roger and Jose meet with Professor Hulbert. They examine the Study Phase work to be accomplished, and decide on how to group the team members to perform that work. Roger and Jose then prepare a list of potential group assignments with Professor Hulbert. Alternative assignments are noted only on Roger's copy. The proposed groups are these:

Roger and Jose
Alice and Joseph
Laura and Kim
Ellery and Manuel

This list is issued as a matrix. The Study Phase activities are listed vertically and the names of the eight team members are listed across the top of the sheet. (See figure 3.3.)

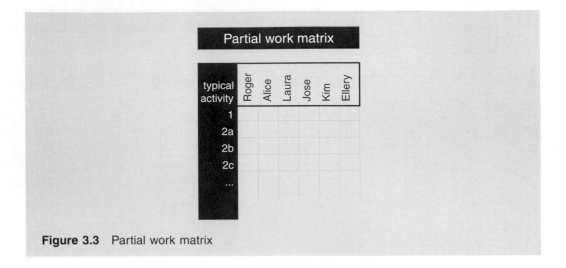

Figure 3.3   Partial work matrix

Activity 1D: Assign Each Group a Set of Tasks to Investigate

Distribute work assignments to each group of team members. Explain to each group how its work relates to that of all the other groups. Allot each team both time and money to perform its part of the Study Phase.

Synthesize your project so the parts can be individually examined and brought together into a coordinated unit. New ideas often occur as each group studies its assignments while noting the work to be performed by the other groups. Periodic review sessions, often referred to as design reviews, are required to ensure that you and all other team members are aware of the progress of the other groups, and how this progress relates to the overall project plan.

PROJECT EXAMPLE

Roger next distributes a list of potential group assignments for discussion. After a few minor changes to satisfy the personalities involved, the groups and their assignments for the remaining activities (beginning on the next pages) in the Study Phase are as follows:

Study Phase Activities	Group
3A	Laura and Kim, with help from Alice
3B	Ellery and Manuel
3C	Laura and Kim
4A	Alice and Joseph
4B	Ellery and Manuel
5A	All team members
5B	Alice and Joseph
6A	Ellery and Manuel
6B	All team members
7A	Roger and Jose
7B	Roger and Jose

The team members all agree to meet each Monday afternoon to review status. Any documents that are to be reviewed by the entire team will be provided to Laura via a computer disk or e-mail by 8 A.M. each Monday morning for printing and distribution that Monday afternoon.

Activity 2: Establish Ground Rules for the Study and Design Phases

Ground rules govern how the group is to function and interact during the next phases and how the work of various members is to be evaluated.

Prepare the output of the Study Phase in a manner that is acceptable to those who will be reviewing the project's progress and its output. In technical matters, a set of rules—often referred to as **ground rules**—governs the conduct, the procedures, and the final products of the Study Phase, as well as the work for the Design Phase (Hosny, Benjamin, and Omurtag, 1994). Ground rules are an established guide for the conduct of the various remaining activities.

When developing ground rules for a *product,* it is important to agree in advance on that product's attributes. Cross (1944, p. 150) notes that product attributes include:

- Utility—performance aspects such as capacity, power, speed, accuracy, or versatility

- Safety—secure, hazard-free operation

- Ease of use—more user friendly (See human factors, section 9.2.)

- Aesthetics—the appearance: color, form, style, surface finish, feel to the touch

- Reliability—freedom from breakdown or malfunction; performance under a variety of environmental conditions

- Maintenance—simple, infrequent, or no maintenance requirements

- Lifetime—except for disposable products, a long lifetime offers good value for the initial purchase price

- Pollution—little or no unpleasant or unwanted byproducts, including noise and heat

Ground-rule examples are given in the case study. When should ground rules be established?

PLANNING AHEAD ▶

Establish the ground rules in advance so that all team members involved will know what the parameters of the projects will be and how their work will be judged. Sometimes you must establish the ground rules during the Conception Phase, modify them during the Study Phase, and further modify them during the early portion of the Design Phase. Thus, team members can examine their work in advance of the formal design reviews to see if they are designing to meet the ground-rule requirements.

As the ground rules are being established, the objectives prepared during the Conception Phase task preparation can be expanded. For smaller projects, most of these objectives are merely stated. For larger projects, where there may be a variety of choices, alternative objectives are prepared, documented, and compared.

One way to assess and compare alternative objectives is known as **weighting** objectives (Cross, 1944, pp. 128–139). He presents a system, with an example, of quantitatively evaluating alternative objectives so you can select the appropriate one. Objectives can include technical and economic factors, user and safety requirements, and so on. **Scores, weights,** or **rankings** may be allocated to each entry, including the qualitative entries.

STUDY

PROJECT EXAMPLE

At a Monday afternoon meeting, Professor Hulbert asks the team to complete a questionnaire that he has developed. It is a list of how the eventual results of the Design Phase of the activity center could be judged. The team members are asked to add their comments for discussion at a meeting to be held Thursday afternoon. (This list, when modified, will become the ground rules for the activity center project.)

At the Thursday afternoon meeting, Professor Hulbert stands at the chalkboard. Team members propose ground rules, and Professor Hulbert writes them on the board. After one hour of intense discussion, the group agrees on a draft of these ground rules.

> The draft of the ground rules for the activity center project is as follows:
>
> - The activity center project must be completed by next August.
> - The cost of the activity center project must not exceed $20 000.
> - The project must be performed primarily by using the skills of the team members.
> - Contractors to be involved must be approved by Professor Hulbert, and their credentials verified by the Northeastern University treasurer's office.
> - All documentation must be prepared as work progresses and be corrected at the end of each phase.
>
> These ground rules will be continually reviewed by all and revised by Alice and Joseph during the remainder of the Study Phase.

Activity 3: Study Solutions to Be Considered for Design

During this activity, you will investigate potential solutions or ways to implement the project and determine the functions to be included in each potential solution. Brainstorming or a similar creative technique (chapter 1) can be applied to devise alternative solutions. You must also determine the risks associated with each potential solution.

Activity 3A: Study Potential Solutions

A key part of the Study Phase occurs when various groups investigate ways to implement the project.

Investigate a variety of solutions to implementing the project. Most ideas can be implemented in more than one way. Where the new solutions represent significant changes in project direction, you should require each group to devise accompanying tasks, schedules, and budgets. The new information, along with the new technical direction, is then evaluated as a part of activity 4B.

Devise approaches that remove the need for technological improvements for projects to be accomplished. If complicated portions of a project concept require a significant improvement in technology, materials, or software, they may be high-risk areas. A simulation may be required. See section 9.3.

Investigate whether potential solutions have been previously devised for other projects. Perhaps no one ever considered applying them to the specific problem involved in your project.

As an example, in electronics, a new type of integrated circuit with a new set of specifications may be needed for the project. You may be asked to prove that this new integrated circuit

- Will be satisfactory
- Can be manufactured in the desired quantity and at a reasonable cost
- Will continue to perform satisfactorily as it ages

The testing effort could require much more investigative time than is available. However, standard accelerated aging processes have been devised and accepted via such organizations as the EIA (Electronics Industries Association). End-of-life characteristics can be approximated through an accelerated aging process so that the time required to force the aging of an integrated circuit is but a fraction of its normal expected life.

Avoid investigating approaches that require excessive expenditure of time and money. Where money is in short supply and time frames require prompt solutions, innovative, untested approaches are usually best avoided. However, some very creative solutions that have been authorized by a team manager have resulted in the savings of large amounts of time and money. For such situations, simpler solutions are often pursued simultaneously along with the more creative solutions that at first seem more risky. Thus, the project will not fail or be terminated because of the risk involved in pursuing only one solution.

Illustration: An electronic component and its testing equipment and procedures were developed by a design team. Considerable testing time was required to verify the performance characteristics of the component. The analysis of the data occurred after the component was shipped to the customer; it required several additional weeks of analysis.

A separate design team of software specialists devised a method to monitor the component as it was being assembled and tested; the acquired data were then analyzed as it was gathered. The results were available to ship with the component, thus saving production-and-shipping time and increasing customer satisfaction.

PROJECT EXAMPLE

Laura and Kim have been assigned to investigate ideas and issues associated with designing an activity center. A second brainstorming session is scheduled. The results are evaluated and the team agrees to explore the following options:

- What locations in Bedford Park are available?

    What are the soil characteristics of the locations?

    What are the building code restrictions?

- What are possible time schedules for construction?
- Have nearby communities developed similar activity centers?

    What do the children using these activity centers think of them?

    What do the teachers, parents, and town and school administrators think of these activity centers?

- How will the goal of meeting children's needs and addressing children's safety issues affect the design?
- How large an activity center can actually be constructed for $20 000?
- What type of activity center should be designed?
- What are the kinds of materials that could be used?

    What types of plumbing materials are available and applicable?

    What are the various types of paints suitable to the activity center exterior and interior, as well as the costs of the paints?

    Will the materials used be available later for repair and maintenance of the center?

- Will landscaping of the grounds be included in the cost?
- What are the possible uses (functions) of the activity center?
- What skills does the team lack that local contractors could supply?

These questions are presented at the next meeting for all team members to analyze. The above list does not change appreciably because it is intentionally vague so it will not become too restrictive during the remainder of the Study Phase. However, during activity 3B below, it is expanded upon as details become available from interviews with individuals experienced in building construction, particularly those contractors and building inspectors who have worked in or near the town of Bedford.

Ellery and Manuel are assigned the task of obtaining copies from the State House Bookstore of the appropriate building, plumbing, and electrical codes that will apply to the activity center structure.

Alice volunteers to investigate alternative foundation designs. She examines textbooks and also interviews Professor Ernest Spencer of the university's Civil Engineering Department regarding the status of concrete foundation designs.

STUDY

As time passes, the actual outcomes of the original project may either be forgotten or may need to be evaluated again. Team members may become so involved with their particular tasks that they forget to consider how their work will meet the needs of the overall project.

Activity 3B: Determine Functions to Be Included in the Solutions

Continually examine the functions of your project with respect to all ongoing tasks. As time passes, the actual outcomes of your original project may either be forgotten or may need to be evaluated again. Team members may become so involved with their particular tasks that they forget to consider how their work will meet the needs of the overall project. The goals of the overall project must be continually examined with respect to the assigned tasks to ensure that tasks are directly and efficiently contributing to the accomplishment of the overall project.

Illustration: In the 1960's, the president of a supermarket chain requested the designers of a major high-tech corporation to devise a system (1) for placing coded labels on grocery-store products and (2) for scanning and reading those labels. The system was to be connected to the cash register. The objective was to decrease customer check-out time.

After spending $1.5M of company funds, the treasurer of the same chain indicated that the more important need was to keep track of inventory leaving the store. Thus, the sold inventory could be replaced overnight on the shelves. At that time, the storage capacity of computing equipment could not handle the additional inventory-tracking requirement, so the design was terminated.

PROJECT EXAMPLE

Ellery and Manuel carefully examine the questions developed in activity 3A. They now convert these questions into functional statements, known as the **system functions.** These functions will guide the Design Phase team.

- The activity center will be located in the northwest corner of Bedford Park.
- The soil consists of 4" to 6" of rocky topsoil with a clay and rock subsoil.
- The building codes that apply are on file at the Bedford Town Hall.
- Construction must start no later than March 1 and must be finished by the end of August because of school term requirements.
- The work will be supervised by the Superintendent of Public Works of the town of Bedford.
- The activity center must be constructed for $20 000. The known variables are these:

    year-round versus summer use

    the number and size of rooms

    the number of floors, including a full or partial basement

    types of roofing, such as slate, wooden, or asphalt shingles

    exterior walls, such as cedar shingles, clapboards, and vinyl or aluminum siding

    exterior trim, such as pine boards, soffits with vents, vinyl, and aluminum

    interior ceilings, such as plaster, drywall, and hung ceilings

    interior walls, such as plaster, drywall, and panels

    interior floors, such as oak, fir, rubber tile, and carpet

    type of heating (such as oil, electricity, and natural gas) if the facility is to be used year-round as an activity center

    landscaping, such as rough grading with sifted existing or purchased loam

    seeding, such as Kentucky blue, rye, and fescue grasses

- The activity center is to provide, for children ages four through ten, recreation opportunities that also develop simple skills such as these:

    sweeping, cleaning, and dusting

    picking up after themselves

manners, such as greeting and thanking others and proper etiquette for eating

sharing and consideration of others during both individual play and team efforts

personal grooming, such as combing hair, brushing teeth, and washing face, ears, neck, and hands

- The team lacks skills and equipment for excavation and foundation construction.

Thus, all the questions in activity 3A that apply to the activity center have been converted to functions that must be considered by the project team.

Activity 3C: Determine Risks and Trade-offs

Develop alternate solutions to the high-risk areas identified during the Conception Phase. Evaluate these solutions by performing a **risk analysis.** During the Conception Phase, the problems and their associated risks were identified. Assess the risks for each proposed alternate solution during this Study Phase. A risk analysis is an analytical technique. It is a quantitative approach used to compare a variety of potential solutions versus their probability of success as a solution. It is a branch of engineering science. For further technical information, see Chase (1974), chapter 9.

Duncan (1996) notes that risk identification consists of determining which risks are likely to affect the project. Each of these risks is then documented and assessed. Risk identification is not a one-time event; it should be performed on a regular basis throughout the project phases. Risk identification is concerned with both opportunities (positive outcomes) and threats (negative outcomes). Checklists are often devised and used for comparison purposes during trade-off studies.

Determine trade-offs among your problem solutions. A **trade-off** is an exchange; it is the sacrificing of one benefit or advantage to gain another benefit or advantage that you consider to be more desirable. List and compare the advantages and disadvantages of each potential solution. A trade-off is a qualitative version of a risk analysis.

In performing a trade-off analysis, you are behaving in a manner similar to a circus tightrope walker. Tightrope walkers sometimes have a safety net under them to catch them if they fall. Projects seldom contain safety nets. There are many competitors—the equivalent of lions and tigers on the circus floor—waiting for your project to falter or fail so they may acquire the work. Therefore, those persons involved in active projects must continually walk the four tightropes noted below.

1. *Staffing: the selection of individuals who can plan and perform a project.* The staff is a combination of (1) highly skilled people who can visualize money-saving changes during implementation and (2) unskilled people who are less costly and need more supervision. Tightrope walking, as related to staffing, is the selection of the optimum combination of these people.

2. *Documentation: the drawings and written descriptions of a project.* It is frequently the desire of some designers to be more specific than is necessary. They overdocument a design to please financial and legal people.

Other designers tend to underdocument in order to save time, provide flexibility, and reduce cost. Tightrope walking, as related to documentation, is the balancing of the amount of documentation desired versus its cost and risks.

3. *Pricing: the cost of labor, material, and services to be involved in a project.* An estimator may prefer to overestimate the cost of various portions of a potential project in order to avoid cost overruns. At the same time, the person who is to manage the project may prefer to underprice a project in order to be competitive and win the contract.

Tightrope walking, as related to pricing, is the weighing of the possibility of being awarded a project that loses money versus losing the project that might have earned a profit. This decision is probably the most critical decision to be faced by the team leader and the team.

*Determine trade-offs among your problem solutions. A **trade-off** is an exchange; it is the sacrificing of one benefit or advantage to gain another benefit or advantage that you consider to be more desirable.*

STUDY

4.   *Cost control: the project manager's concern once the project is funded.* It is the responsibility of the project manager to watch every penny that is being spent and how it is planned to be spent. Some managers may not observe that many dollars are being wasted while only a few pennies are being saved. Other managers allow some flexibility in group expenditures so that the ingenuity of the team members may be used to keep total expenditures to a minimum.

Tightrope walking, as related to cost control, is the balancing of authorized expenditures versus the work to be performed. It is a nontechnical activity and depends greatly upon the attitudes of the manager and team members concerning money.

Select the best option for each of the above categories. Examine the result of combining these options. Verify that this combination is the optimum combination. This effort is known as performing a trade-off among the various options. The selection of the optimum combination of these options continues to be more of an art than a science.

PROJECT EXAMPLE

Laura and Kim carefully examine the activity 3B list of system functions. They assess the potential risks associated with each proposed system function.

- What are the problems with locating the activity center in the northwest corner of Bedford Park with respect to water, sewage, electricity, or natural gas access?

- Where is electrical power available during construction?

- Are there available locations within Bedford Park for temporary storage and further processing of the excavated soil?

- Which portions of the Bedford Building Codes apply to this project?

- What are the ranges of environmental conditions, such as temperature, humidity, wind, and noise levels, that occur in this area for either year-round or only summer-use considerations?

- What are the cost comparisons for constructing more than one floor level and the effect upon construction and heating costs?

- Which building materials are locally available—including their prices—for the roof, exterior walls, interior walls, ceilings, and floors?

- Should landscaping be included as a part of the project, or should this work be performed by the Bedford Public Works Department?

- What local companies and employees might be willing to contribute materials, labor, or only advice and guidance?

- What foundation construction techniques are applicable?

A trade-off study is sometimes required. As an example, concrete floors are of a thickness that may require steel mesh within the concrete for structural support. If the concrete floor is 20 percent thicker than is normally used in this type of building design, then the internal steel mesh (or reinforcing rods) will not be required. Both approaches must be evaluated in terms of cost and risk.

A trade-off study is performed by Kim Wong, assisted by a fourth-year technology student. Test materials are prepared, and stress testing and other measurements are performed to determine the resistance of the floor to cracking. That fourth-year student seems more interested in the results of his tests than in the activity center schedule. In order to get him to complete the tests on time, Kim has to continually remind him that the activity center schedule must be followed.

Activity 4: Gather and Evaluate Information

During this activity, you will contact potential vendors for their technical and financial input, and then evaluate design alternatives. Along with this evaluation, you will be refining your cost estimates.

Activity 4A: Contact Potential Vendors for Their Input

The input of vendors or subcontractors can be invaluable in guiding a project. If they expect to be awarded work as a result of their participation, they may offer advice at no cost.

Identify other persons or organizations who have the additional skills and abilities needed to assist in the performance of your project. They may be **vendors** or **subcontractors.** Their contributions may prove invaluable, particularly if you involve them in the Study Phase effort. They will then feel they are a part of your team. They may frequently offer excellent suggestions based on their experience. If they expect that a subcontract may be offered to them, they may also provide the Study Phase input at no cost to your project.

PROJECT EXAMPLE

Alice and Joseph are given a list of the skills of the team. They then develop a list of the skills necessary to construct the various designs of the activity center. They are now able to identify those skills and the equipment that will be necessary to both design and construct the activity center.

Alice's uncle offers to work with the Northeastern University faculty in reviewing the activity center design as it progresses. Professor Hulbert identifies and urges other faculty members to act as design reviewers. He recommends that the students study the architectural and construction options given in Muller, Fausett, and Grau (1999). For the Study Phase, Joseph contacts several local contractors who can either loan or rent the necessary earth-moving equipment. He also obtains construction cost estimates from them. Alice locates a contractor who will assist in the construction of the concrete foundation, if that is the design path chosen. No commitments are offered at this time because the design path has yet to be selected.

STUDY

Activity 4B: Evaluate Design Alternatives and Refine Estimates

Cost estimates need to be reviewed to ensure that pricing data are up-to-date and realistic. This activity is primarily a technical activity, with some emphasis on cost and schedule changes.

Examine the outcomes of your previous Study Phase activities. Value analysis may be helpful. (See appendix A.) Compare all of the proposed solutions to your project with one another. Prepare comparison documentation in one standard format in preparation for the work of activity 5A on page 72. The designers will evaluate these alternatives. If cost estimators are involved, they will provide information that will guide the designers. Review cost estimates to ensure that pricing data are up-to-date and realistic. This activity is primarily a technical activity, with some emphasis on cost and schedule changes.

PROJECT EXAMPLE

At their next Monday afternoon meeting, all activity center design alternatives are examined and evaluated. The ground rules are used as the judgment vehicle for both the team members and the work they accomplished. On the board, a matrix is developed that compares each alternative solution with the previously devised ground rules. It is noted that no one has evaluated the effect of the applicable building codes upon each proposed design solution.

Ellery and Manuel obtain a copy of the local and state building codes, including the plumbing and electrical codes. They discover that the various applicable codes originated with the Council of American Building Officials. The applicable codes are the following:

- National Electric Code
- CABO One- and Two-Family Dwelling Code
- BOCA National Building, Plumbing, and Mechanical Codes

They extract the following information that may be of value to the designers:

- Foundation drainage grade: $\geq 6''$ per $10'$
- Stud drilling & notching $\leq 25\%$ of width; on-load bearing $\leq 40\%$ of width; doubled studs $\leq 60\%$ of width
- Window air filtration < 0.5 CFM per linear ft with $\Delta p = 1.56$ psf (CFM = cubic ft per minute; psf = pounds per square ft)
- Handicapped access ramp, slope 1:8 maximum and 1:12 minimum; landing $\geq 3' \times 3'$; guard rails $\geq 36''$ high
- Stairway $\geq 42''$ wide; handrails $\leq 34''$ high; landings $> 3' \times 3'$
- Hallway width $\geq 36''$
- Smoke detectors: all floors—for new construction, permanently wired to 115V

Manuel is assigned to devise a schedule for completing the Study Phase work and to prepare the cost spreadsheet for completing that work.

Activity 5: Select and Plan One Solution

During this activity, you will select one solution for your project, and then plan the Design and Implementation Phases based upon your solution.

Activity 5A: Select a Solution

For selection of solutions or ways to implement a project, the following criteria should be considered:

- *Complexity and cost of the solution*
- *Time to implement solution*
- *Preferences of clients*
- *Availability of materials*

Select the project solution from your previously devised options and design alternatives by comparing the following factors:

- Complexity (technically complicated and intricate)
- Costs (labor and material)
- Urgency (time to investigate versus limited time available)
- Preferences of clients
- Availability of materials

Such decisions are often nontechnical ones.

Your options represent choices that must be weighed both separately and together because they interact, as noted in figure 3.4.

Decision making is a continuous effort of analyzing, assessing, and choosing. Technical information is gathered and compared. Financial information may fluctuate rapidly or may change slowly. The perceptions of your client may be different from your team's preferred technical solution—the client's input must also be considered. The project manager must "smooth the ripples" of change so that team members do not frequently proceed in different directions. During the decision-making process, individuals may become confused and frustrated and no longer perform as a team!

When you are involved in large projects or programs, a formal mathematical or statistical model may be needed for making decisions. See chapter 9. For smaller projects, you can develop a matrix of information in the form of a decision tree for comparison and consideration purposes.

Figure 3.4 Decision basis for project choices

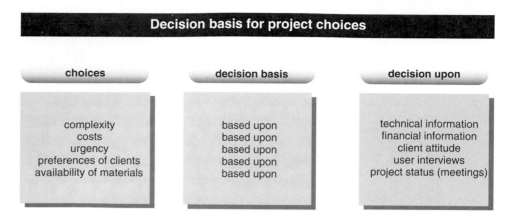

Decision basis for project choices

choices	decision basis	decision upon
complexity costs urgency preferences of clients availability of materials	based upon based upon based upon based upon based upon	technical information financial information client attitude user interviews project status (meetings)

PROJECT EXAMPLE

All team members participate in devising the design direction for a project. This may be a day-long process. By the end of the day, the team members can produce a sketch of the activity center they propose.

All team members participate in devising the design direction, which is to be followed during the next phase of the activity center project. The decisions require so much uninterrupted time that an entire day is set aside for a solution selection. A matrix of information, in the form of a decision tree (figure 3.5), is prepared by Ellery and Manuel as a part of performing activity 4B.

STUDY

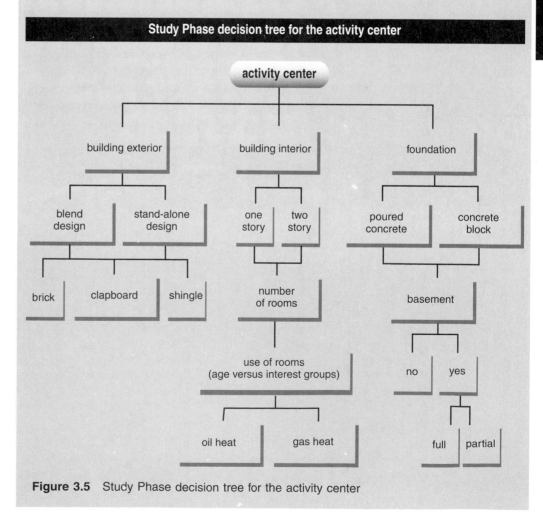

Study Phase decision tree for the activity center

Figure 3.5   Study Phase decision tree for the activity center

Several questions (see chapter 2, activity 2F) are considered during this session:

- Will the activity center be constructed to blend in with other local structures and the land, or should it stand out for reasons of its uniqueness?
- Should the exterior finish be brick, clapboard, or shingle?
- Will it be a one-story or two-story activity center?
- How many rooms should there be within the activity center?
- What uses are to be considered for each room?
- Should the activity center rooms be designed for special age or interest groups?
- Should the central heating system be oil- or gas-fired?
- Will the activity center require a basement?
- Should the foundation consist of concrete block or poured concrete?
- If there is a basement, should it be a partial or full basement?

The work of Kim regarding foundation and concrete floors is examined first because it affects the design choices for the remainder of the activity center. It is decided to use the thicker 6″ concrete floor; it will cost less and will not crack under normal conditions of use.

The decisions agreed on at the meeting that affect the Design Phase (covered in chapter 4) are the following:

- The activity center will be constructed for year-round use.
- It will have a first floor with three rooms, one toilet per room.
- The center will have a basement that can be used for a furnace and for storage of toys that are used outdoors.
- The foundation and first-floor walls will be constructed to allow for later roof raising and the addition of a second floor.
- Roofing will be asphalt shingles.
- Exterior walls will be clapboards.

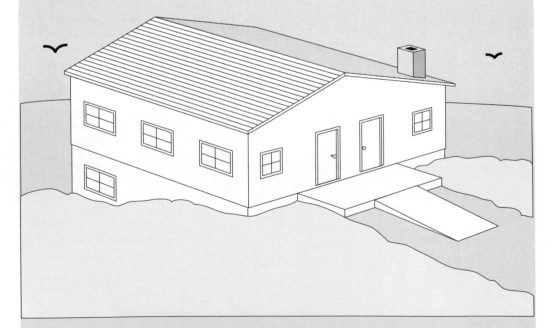

Figure 3.6   Preliminary sketch of activity center design

- Exterior trim will be pine boards.
- Interior ceilings will be drywall.
- Interior walls will be drywall.
- Interior floors will be rubber tile throughout.
- The type of heat will be natural gas.
- Landscaping will use existing loam $\geq 4''$ deep for finished grading.
- The seed will be supplied by the Town Department of Public Works.

Professor Hulbert reminds the team members to compare these decisions with the ground rules devised in activity 2. A preliminary sketch is shown in figure 3.6.

Activity 5B: Plan the Design and Implementation Phases

The task descriptions, costs, and schedules for the project can now contain more detail, as the project moves nearer to the Design Phase.

Review and revise the Conception Phase plans for the Design and Implementation Phases. Include applicable results from the Value Analysis (see appendix A). Ensure that the revised plans remain realistic with regard to task description content, schedules, and budgets.

For the task descriptions, state the content of each task in enough detail that there are no misunderstandings among the potential workers and managers. Task descriptions define the content that is indicated in the task title. See appendix B.

More detail must be added to the schedules devised during the Conception Phase. Potential vendor and subcontractor plans must also be included in the revised schedules. Schedule headings should coincide with the task headings. Often the titles for the task descriptions are used as headings in schedules so that schedules and tasks coincide. They are often shortened so they fit in the space available.

Budgets must be updated with justification provided in the form of price quotations from both vendors and materials suppliers. These budgets should include headings that coincide with the task and schedule headings.

PROJECT EXAMPLE

PLANNING AHEAD▶

Alice and Joseph work closely with Ellery as he writes the task descriptions, and with Manuel, who is revising the schedules and budgets. They need to identify the work to be accomplished during the Design Phase, as well as the human skills and resources needed. They must also estimate the human skills and resources needed for the Implementation Phase. For these two phases, they identify either team members or vendors who can provide time and skills, facilities, equipment, and computation support. The task descriptions, schedules, and budgets are now reviewed by all the team members and revised as necessary. Roger prepares a plot plan (figure 3.7). He has to remind them that, even though vacation week is approaching, there is still work to be done.

Professor Hulbert notes that this is the time to learn how to prepare the following documents.

- The Letter of Transmittal, which for the activity center will be a contract letter
- A Design Specification for the activity center, which describes each item to be constructed
- The working and detail drawings, produced with a CAD program

STUDY

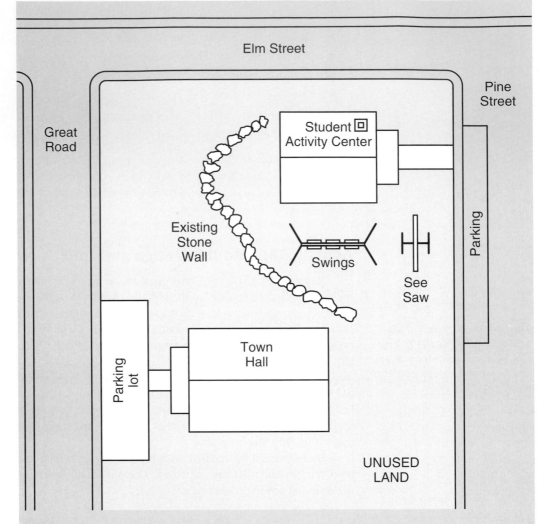

Figure 3.7   Activity Center Plot Plan

He requests that each team member reexamine the requirements of the Design Phase (chapter 4) and indicate where they can either apply their existing skills or can quickly learn new skills.

Roger Stein is overwhelmed by the variety of work to be performed in the Design Phase and wonders where the team will locate the people with the necessary skills.

Alice D'Annolfo believes that she can write most of the Design Specification because of her experience in working with her father. However, she is not comfortable with the legal language involved.

Laura Drake would like to learn how to use a CAD program to devise working drawings, but she worries about how many she can produce during the eight weeks of the Design Phase.

Jose Hernandez prefers to work on revising material costs during the Design Phase, as he is going to have an increased class load next term.

Kim Wong wants to work on foundation and concrete specifications and to continue to refine the cost estimates.

Ellery Kistler agrees to help the other students learn the CAD program.

Manuel Cardoza prefers to concentrate on applying the CAD program.

Joseph O'Neil wants to learn the CAD program and to assist in refining the lumber and related estimates.

Professor Hulbert offers to contact the University Law School for the assistance of one professor or lecturer who has had extensive experience in either writing or reviewing contracts and specifications for both private and commercial buildings.

Activity 6: Document the Selected Solution

As a part of your documentation, you will refine your tasks, schedules, and budgets for the Design and Implementation Phases. You will also prepare a description of your proposed solution and the rationale for its choice.

Documentation becomes more crucial as more people are involved in a project. Descriptions of tasks and subtasks must be easily understood by the Design Phase designers and the Implementation Phase builders.

Activity 6A: Refine the Task, Schedule, and Budget Documents

Individual team members, and groups of team members, have developed the details regarding the selection of the Study Phase project solution and the Design Phase options. Refine your tasks, schedules, and budgets for the Design and Implementation Phases. (See chapter 7.) Note within your documentation that the Implementation Phase documents are best estimates based on available information.

Rearrange your tasks and subtasks so they are more easily understood by the Design Phase designers and the Implementation Phase builders. Information that you gather and examine during this Study Phase may cause new tasks to replace tasks previously documented. Thus, your task descriptions, schedules, and costs for the next phases may change in both content and sequence.

PROJECT EXAMPLE

Ellery and Manuel realize that they must work with the other teams to determine design directions and details investigated by these teams and to determine the various associated costs. They decide not to write task descriptions except for those areas that are firm. They discover that it is wiser to invest their time pursuing the costs for firm items and to estimate the accompanying schedule changes rather than to write all the task descriptions at one time.

By the middle of December, with exams over and the vacation week very near, Manuel has devised a modular computer program that can process the inserted changes in cost and schedule (chapters 7 and 8) and display these changes without the need to print another copy. This is of tremendous assistance to the team in rapidly examining the effect of suggestions for modifications in design.

As the Study Phase nears completion, Ellery and Manuel realize that they cannot write all the task descriptions for the Design Phase within the time remaining. Joseph assists Ellery and Manuel in writing these descriptions. The entire team agrees to review the results each evening prior to vacation.

The task descriptions change. Various subtasks are regrouped so they "make more sense" to the ultimate designers and builders. New tasks are added. Sometimes tasks are eliminated. (It should be noted that even these tasks, schedules, and budgets may change again during the Design Phase.) Compare these two lists of tasks:

Tasks from the Conception Phase	Tasks from the Study Phase
Prepare Site	Prepare Site
	Cut and Remove Brush
	Remove Protruding Ledge
Excavate Site	Excavate Site
	Define Foundation Corners
Erect Foundation	Erect Foundation
	Waterproof Exterior Foundation Walls
	Backfill Completed Foundation
	Remove Excess Dirt
Cap Foundation	Cap Foundation
Erect Rough Framing and Sheathing	Erect Rough Framing and Sheathing
	Erect Exterior and Interior Studs
	Erect Roof Rafters
	Board in Walls and Roof
Shingle Roof	Shingle Roof
Install Windows	Install Windows
Install Heating	Install Heating
	Install External Doors
Finish Exterior	Finish Exterior
	Apply Prime Coat of Paint to Exterior
Install Sheetrock™	Install Sheetrock™
Plaster Interior	Plaster Interior
Lay Floors	Lay Floors
	Install Vapor Barrier
	Install Flooring
Install Interior Finish	Install Interior Finish (Doors, Windows, Baseboards)
	Install Stair Treads and Risers
Paint Interior Finish	Paint Interior Finish (Ceilings, Walls, Trim, Floors)

Once the above task details are established, the schedule of tasks is revised. The new schedule shown in figure 3.8 should be compared with that in figure 2.8 in chapter 2 (the upper schedule in figure 3.8). The lower schedule in figure 3.8 has been modified to reflect the knowledge acquired during the Study Phase.

Similarly, the cost-information details are refined as the tasks and schedules are developed in greater detail. Figure 3.9 should be compared with figure 2.9 in chapter 2 (the upper estimate in figure 3.9). Figure 3.9 has been modified to reflect the knowledge acquired during the Study Phase.

Roger asks Professor Hulbert if the tasks should be numbered so it would be easier to follow them as the team progresses into the Design Phase. Professor Hulbert considers the question and then decides they should wait because the changes in task sequence and content during the Design Phase will most likely

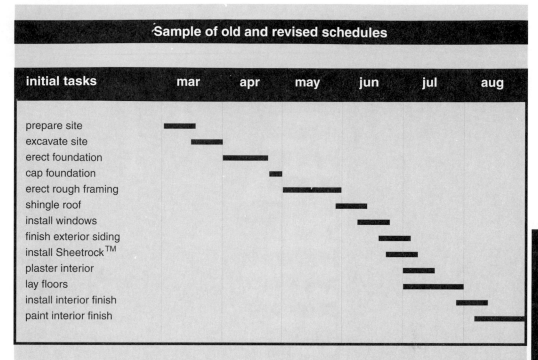

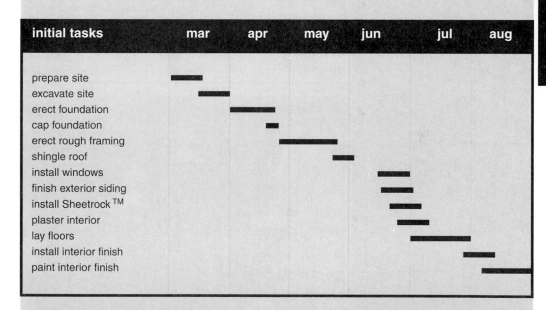

Figure 3.8   Sample of old and revised schedules

be significant. Numbering the initial tasks at this point could cause confusion because the numbers will change when the design is in progress. He suggests that they should use letters at this time so they can more easily compare the Study Phase lettered initial tasks with the Design Phase numbered tasks.

The final task descriptions, schedule, and budget are refined. Professor Hulbert notes that there seem to be no *significant* changes since these budgets, tasks, and schedules were developed during the Conception Phase.

Ellery Kistler asks for the opportunity to write the details of the task descriptions so that he can extend his English writing abilities to technical English. (See appendix B.) Two examples of his writings are given below to help you understand the concept.

Sample of old and revised cost estimates

initial tasks	mar	apr	may	jun	jul	aug	totals
prepare site	600						600
excavate site	1400						1400
erect foundation		1700					1700
cap foundation		400					400
erect rough framing			4800				4800
shingle roof				1000			1000
install windows				1800			1800
finish exterior siding				400	400		800
install Sheetrock™				300	500		800
plaster interior					600		600
lay floors					2200		2200
install interior finish					600	1200	1800
paint exterior finish						1300	1300
totals	2000	2100	4800	3500	4300	2500	19 200

initial tasks	mar	apr	may	jun	jul	aug	totals
prepare site	600						600
excavate site	1400						1400
erect foundation		1700					1700
cap foundation		400					400
erect rough framing			4800				4800
shingle roof				1000			1000
install windows				1800			1800
install heating							0
install exterior doors					400		400
finish exterior siding				400	400		800
install Sheetrock™				300	500		800
plaster interior					600		600
lay floors					2200		2200
install interior finish					600	1200	1800
paint interior finish						1300	1300
totals	2000	2100	4800	3500	4700	2500	19 600

Figure 3.9   Sample of old and revised cost estimates

Prepare Site

Survey the site, with Registered Land Surveyors or Civil Engineers, to establish grade levels of the basement floor. Place a stake at each corner of the foundation with "depth of cut" noted on each stake. Install batter boards at each corner and mark them with the location for each foundation wall.

Excavate Site

Excavate the basement and foundation footings, with a minimum of 1' clearance beyond the exterior walls of the foundation. If interfering ledge outcroppings are discovered, then drill with 2" diameter rock drill bits to a depth of 1' below the designed finished-floor level, and blast to remove that ledge.

Note that the first task description (Prepare Site) directly leads to the second task description (Excavate Site). All other task descriptions connect to at least one previous task description and lead to at least one other task description.

The entire set of task descriptions, schedules, and budgets is again reviewed; the review lasts an entire afternoon.

Activity 6B: Document the Proposed Solution and Rationale

The documents for the Study Phase are similar to those from the Conception Phase, except they are much more detailed.

The documents for the Study Phase are the following:

- The Project Description with design options
- The Design Specification
- The Design Description with preliminary sketches

The content of each of these documents is described in section 3.3 on page 83. Preliminary sketches from the solution section process are included. (See activity 5A.) The Design Description should also avoid assumptions that direct the Design Phase team toward considering only one design.

PROJECT EXAMPLE

Roger assigns each member a portion of the following documents to be written:

- The Project Description for the activity center, with design options
- The Design Specification
- The Design Description with preliminary sketches

Roger notes that he has been reviewing portions of these documents for two weeks. He states that it is time to write the missing paragraphs and sections and to read each document for its content and the flow of ideas. Here are the assignments for preparation of the documents:

- Roger and Jose turn over the Project Description for the activity center to Alice for review.
- Alice and Ellery turn over the Design Specification to Laura for review.
- Roger agrees to write the Design Description.
- Ellery agrees to prepare the preliminary sketches of the activity center.

They plan for one last meeting before the week-long school vacation.

STUDY

Activity 7: Verify and Obtain Funding

During this activity, you will first verify the actual availability of the funding identified during the Conception Phase, and then request the Design Phase funding in an official manner.

Activity 7A: Verify Actual Availability of Funding

The continuing availability of funds from sources needs to be constantly monitored, so that the project can obtain funds when they are needed.

Confirm periodically how much funding continues to be available for your selected project. Determine the latest time that it will be available. During the Study Phase, this activity should be performed by the team manager or a small group of individuals from your Conception Phase team. Why? Because funding may be reassigned to other financially needy groups or organizations while your Study Phase team is still working to devise a project solution that will satisfy both your team and the funding source.

PROJECT EXAMPLE

Roger and Jose discover that the Commonwealth of Massachusetts has reduced the $15 000 in funding to $10 000 because of state budget changes. Miss Kewer's $5 000 continues to be available. Thus, the Study Phase team must either revise its options and construct an activity center for $15 000 or else seek an additional $5 000 elsewhere. Roger, Jose, and Professor Hulbert investigate other sources that would be willing to contribute the needed additional $5 000.

The current status of the project can be summarized as follows:

- The activity center lacks $5 000.
- Other funding sources are not yet committed to contributing to the activity center.
- The skills of team members remain applicable to the various designs for the activity center.

Even though an additional $5 000 needs to be raised to construct the $20 000 activity center, a $20 000 activity center seems to be the preferred project level.

Although the team members seem frustrated and discouraged, Professor Hulbert explains that such a funding problem is not unusual. He instructs the team to keep the $15 000 as a lower bound for the Design Phase while he, Roger, and Jose investigate other options for obtaining more funding.

Activity 7B: Obtain Design Phase Funding

Provide a presentation to the previously identified funding source. Verify that your project can be funded promptly. It is customary for funding sources to retain control over their funds until their personnel have reviewed the results of your Study Phase. Now revise those portions of the project as requested by the personnel from the funding source. If it is necessary to have no lost time between the Study and Design phases, then provide a preliminary presentation to the funding source prior to the completion of your Study Phase work.

Funding authorization may occur in one of two ways. Such authorization is usually in the form of a letter. For foundations, a check for the agreed amount is often included with the letter of authorization. This, in effect, pays for the work in advance. For government agencies or within the parent organization of the project team, a purchase order may be issued to authorize the project to proceed.

In conjunction with Professor Hulbert, Roger obtains the additional no-cost services of vendors for the Design Phase, who later hope to participate in the construction of the activity center. Many of these are contractors who have worked together before in the construction of both private and industrial projects in the Bedford area. They are instructed by Professor Hulbert to be careful *not* to perform work that should be a learning experience for the team.

The team members agree to meet on Tuesday afternoon following their last classes to see if they have discovered any significant errors in the documentation of the Study Phase and the proposed solutions. All the vendors agree to meet at Northeastern University on the Wednesday evening following the year-end school vacation to review the Study Phase documentation.

An important note: The group of Professor Hulbert, Roger Stein, and Jose Hernandez realize that all of the work to date will be wasted if the necessary additional $5 000 is not acquired soon. Otherwise, the Design Phase must limit implementation costs to $15 000. They agree to talk to friends and family during the University vacation for ideas relating to, and sources for, the additional $5 000.

It is late December. Now it is time for Roger to terminate the Study Phase so that they can proceed into the Design Phase when they return and so that all team members can enjoy a well-deserved one-week vacation.

SECTION 3.3   STUDY PHASE DOCUMENTS

During the Study Phase, the ideas developed and documented during the Conception Phase are examined in more detail. Using the Project Description with Artist's Concept and Preliminary Specification as guides, members of the project team proceed with further investigation. In order to more accurately estimate the tasks, schedules, and costs needed to design and construct the selected project, team members propose and pursue a variety of solutions to the selected project.

As the Study Phase progresses, more ideas are circulated via papers and memoranda; more meetings are held to compare and evaluate the various proposed solutions, and more meeting reports (section 8.2) are written. These memoranda and meeting reports eventually lead to three Study Phase documents. The documents completed by the end of the Study Phase are the following:

1. The Project Description (with design options): explanatory documents
2. The Design Specifications: the second phase of specification development
3. The Design Description with preliminary sketches: more detailed drawings and plans

The content of each of these documents is described below. The flow from the Conception Phase documents to the Study Phase documents is shown in figure 3.10.

The **Project Description** consists of the following:

1. A description of the potential project solutions that were explored
2. The attributes of each explored potential project solution
3. The tasks, schedule, and budget for the selected project solution
4. Justification for the selected project solution

This document is often a modified and expanded version of the Project Selection Rationale document from the Conception Phase. This new document typically consists of less than fifty pages for a modest-size project. For a large, complicated program, this document may consist of several volumes of applicable information.

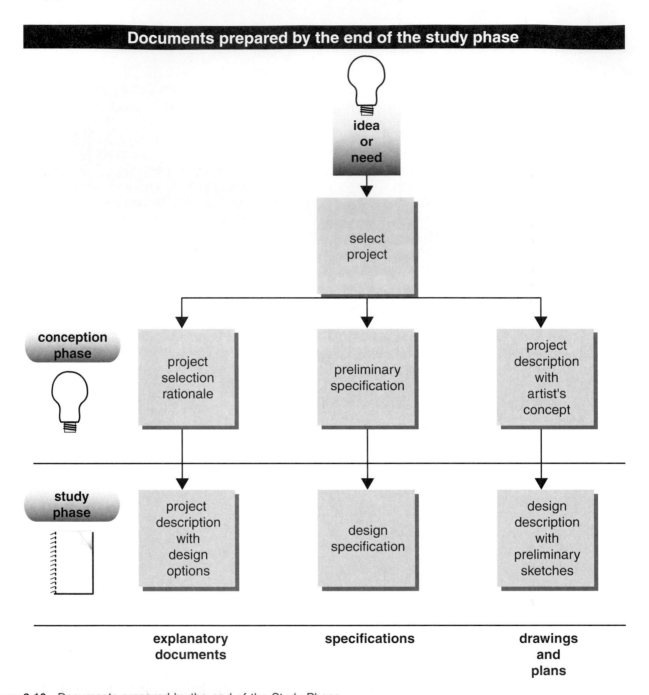

Figure 3.10   Documents prepared by the end of the Study Phase

*The documents from the Study Phase guide the Design Phase. The **Design Specification** document establishes the constraints for the Design Phase.*

The **Design Specification** document establishes the constraints for the Design Phase. It evolves from the Preliminary Specification. The content of the Design Specification is described in chapter 8. The Design Specification must reflect only those functions desired for the selected solution. However, it should avoid directing the Design Phase team toward only one design approach.

The **Design Description** should contain the following:

1. Justification for the project selection developed from the Project Description document
2. A semitechnical description that nontechnical persons can comprehend
3. Estimated schedules per phase
4. Estimated costs per phase

5. Directions for the Design Phase

6. Staffing skills required for the Design Phase

7. Resources needed for the Design Phase

Preliminary sketches for proposed solutions that evolve from the Study Phase should be included. However, the Design Description must avoid guidance that directs the team toward only one solution during the Design Phase.

SECTION 3.4

The Study Phase is considered complete when the documentation has been written, reviewed, and revised as recommended during its review. The review team should include individuals from your funding source.

The Study and Design Phases may overlap as some members begin to prepare more detailed designs for the project.

DEFINING PHASE COMPLETION

The Study Phase is considered complete when the documentation has been written, reviewed, and revised as recommended during its review. The review team should include individuals from your funding source. Wherever it is possible, include as reviewers some experts in the field who will not be responsible for the funding of the later phases. Why? Because they have not been "close" to the project on a daily basis and may notice problems that have not been considered.

Someone within your organization needs to monitor the total project costs on a continuing basis. This is necessary for all phases of a project because you do not want to overspend, you must stay on schedule, and you want your total effort to be performed according to the specifications previously developed.

The completed documents for the Study Phase consist of an expansion and revision of documents originally devised during the Conception Phase. They are the following:

• The Project Description with design options, which has evolved from the Project Selection Rationale and includes the refined cost estimates

• The Design Specification, which has evolved from the Preliminary Specification

• The Design Description with preliminary sketches, which has evolved from the Project Description (with the artist's concept)

Often, these three documents are initiated during the beginning of the Study Phase. They gradually evolve into their final content by the end of the Study Phase.

The Study and Design Phases may overlap because of the following:

• Some members of the group are still investigating questions remaining from the Conception Phase

• Other group members are starting to sketch potential designs and select the materials needed for the Design Phase

• Other group members are starting to prepare the Design Specification

• Other group members are writing portions of the Production Specification

Decide when the overall Study Phase will end. Realize that more information may need to be gathered as the design is being developed.

Revise the entire project if very unusual information suddenly becomes available. However, such a revision may discourage the funding source. Thus, a significant project revision should be avoided except during extreme situations. Personnel from the funding source should be kept aware of any significant changes, as well as of the reasons for those changes.

Increase the detail of the tasks and task descriptions devised during the Study Phase as more information is gathered. For some projects, the tasks may also be resequenced. The first three phases of a project may interact as shown in figure 3.11.

Start at the top of figure 3.11 where the concept is determined. For a Yes answer to Concept Determined, you may proceed to the Study Phase. However, if the answer to Concept Determined is No, then you must return to the Conception Phase and revise the appropriate activities. Once the project is selected during the Study Phase, you may proceed to the Design Phase.

Figure 3.11 Interaction of the first three phases of a project

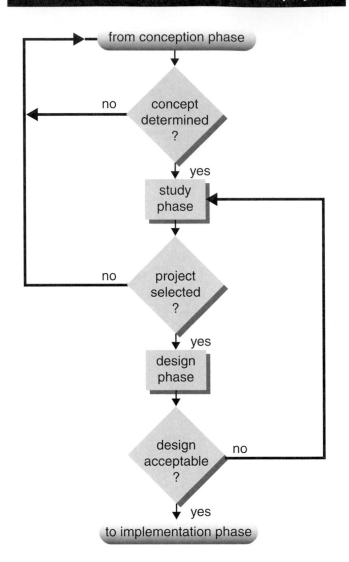

Interaction of the first three phases of a project

There will be times when, during the Design Phase, the design may not be acceptable. If the answer to Design Acceptable is No, then you must return to the Study Phase and revise the appropriate activities.

Continue to walk the tightrope of deciding when a phase is complete. Note any problems that linger from the Study Phase that could adversely affect your Design Phase. Assign problem-solving responsibility to a small but proficient team to rapidly resolve the remaining problems.

CHAPTER OBJECTIVES SUMMARY

Now that you have finished this chapter, you should be able to:

1. Describe the purpose and goal of the Study Phase of a project.

2. Explain the tasks involved in the Study Phase of a project and the activities or steps through which these tasks are performed.

3. Describe the process of reviewing the Conception Phase and planning for the Study Phase.

4. Describe the process of utilizing skills of team members to complete the Study Phase and managing team members to achieve the overall goals of the project.

5. Explain what ground rules for a project are and how they are established.

6. Explain how to investigate and study solutions for ways to implement the project and to evaluate the risks and trade-offs involved with various solutions.

7. Describe how to select among design alternatives as a group process.

8. Discuss relationships with funding sources and potential vendors during the Study Phase.

9. Describe the documents prepared during the Study Phase and how they evolve from the Conception Phase documents.

EXERCISES

3.1   Revise the Purpose and Goal statement given in section 3.1 to match each of your class-project options.

3.2   For your revised Purpose and Goal statement, review your Conception Phase results and organize your Study Phase work. Establish alternative solution requirements and accompanying constraints.

3.3   Carefully examine the activities listed and described in section 3.2; adapt them to each of your options for a class project.

3.4   Write and review the ground rules (criteria) for judging your Study and Design Phase results. Verify that these ground rules are compatible with the expectations of your client and funding sources.

3.5   Review your Conception Phase effort with all team members and other interested parties. Develop the Study Phase content details with all individuals concerned.

3.6   Organize the team into small groups based on their skills, compatibility, and availability. Verify or revise your choice of team leader. Assign each small group a set of tasks to investigate. Develop alternative solutions that satisfy the devised criteria.

3.7   Identify and evaluate risks and trade-offs for each of your proposed alternative solutions.

3.8   Examine the effort required to revise the Conception Phase tasks, schedules, and budgets. Compare this effort with the time and personnel available. Discuss how you might perform this effort without a significant loss of end-product quality.

3.9   Study potential solutions to the selected class project. Develop a list of functions and investigate those forms (solutions) that will satisfy those functions.

3.10 Determine the necessary risks and trade-offs required to implement the project solutions. Devise new tasks, schedules, and budgets where necessary. Contact potential vendors for their technical and financial input.

3.11 Request that the previously identified judges evaluate your proposed solutions. Evaluate accepted design alternatives and refine cost estimates where necessary. Select a project solution.

3.12 Verify the continued existence, actual availability, and method of receiving funding.

3.13 Refine all task, schedule, and budget documents for the Design and Implementation Phases. Document the proposed project solution and the reasons for its choice.

3.14 Prepare documentation describing your proposed design alternatives.

3.15 Obtain Design Phase funding and terminate the Study Phase.

3.16 Describe how you would determine the optimum time to terminate the Study Phase.

The Systematic Approach

The Conception Phase

The Study Phase

The Design Phase

The Implementation Phase

chapter **4**

1. Systematic Approach

2. Conception Phase

3. Study Phase

4. Design Phase ▶

Design Phase Purpose and Goal

Design Phase Activities

Design Phase Documents

Defining Phase Completion

5. Implementation Phase

1. What are the purpose and goal of the Design Phase of a project?

2. What tasks are involved in the Design Phase of a project, and what are the activities or steps through which these tasks are performed?

3. What is the role of a manager in the Design Phase?

4. How can a team be organized for the Design Phase and tasks be assigned to individual team members?

5. What is a typical schedule for the Design Phase of a project?

6. What human factors are involved in designing a product?

7. How are function and form linked in developing design plans?

8. What are make-or-buy decisions, and how are they decided?

9. What documents should be prepared for the Design Phase?

10. Who should participate in reviewing and approving the documents from the Design Phase?

11. What are the uses of the documents prepared in the Design Phase?

12. How can it be determined when the Design Phase is completed?

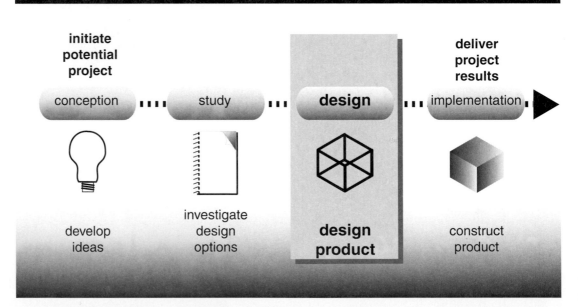

...[D]esign is a polishing job. Much as the potter's wheel is used to mold and shape the first-hewn piece of clay, so the system-design process, through repetition of its steps, molds and shapes the final system.

H. H. Goode and R. E. Machol, 1957

This chapter emphasizes the development of the plans for the project introduced in Chapter 2 and expanded in Chapter 3.

SECTION 4.1   DESIGN PHASE PURPOSE AND GOAL

The Design Phase of a project develops the working plans.

The **purpose** of the Design Phase is to develop the details for the working plans for your proposed project solution. The **goal** of the Design Phase is to prepare a complete set of working plans for the selected design. During this phase, it will be necessary to continually review the working plans for the selected design, including cost estimates, as the design evolves.

At the end of the Design Phase you should know how to accomplish the following:

- Organize your teams into groups for performing design tasks.
- Study potential implementation solutions.
- Develop a design that will satisfy you, your funding source, and the users of your product.
- Refine your previously estimated task, schedule, and budget documents so that they are well developed for the Implementation Phase.
- Document your proposed project design and the reasons for its choice.
- Verify that both your customer and your funding source will accept the Implementation Phase solution.
- Verify availability of funding for the Implementation Phase.

SECTION 4.2   DESIGN PHASE ACTIVITIES

The activities for the Design Phase consist of the following:

Activity 1: Organize the Design Phase Work

   1A: Review the Study Phase effort.

   1B: Develop Design Phase details.

   1C: Organize teams and assign design tasks.

Activity 2: Select the Solution Details to Be Implemented

   2A: Study potential implementation solutions.

   2B: Contact vendors for make-or-buy information.

   2C: Determine make-or-buy decisions.

Activity 3: Document the Design Solution

   3A: Prepare the Production Specification.

   3B: Prepare the Working and Detail Drawings.

   3C: Refine tasks, schedules, and budgets.

   3D: Document the design and reasons for its choice.

Activity 4: Prepare for the Implementation Phase Work

   4A: Review the Design Phase results.

   4B: Verify availability of Implementation Phase funds.

   4C: Plan the Implementation Phase.

Activity 5: Obtain Implementation Phase Funding

As is true for the Conception and Study Phases, the purpose and goal you define for the Design Phase determine the activities that you are to accomplish. The sequence of activities given in this chapter may change from one project to another. However, all the activities must be accomplished before the Design Phase is complete.

As was true during the Study Phase, there will be (design) risks involved with each alternative solution you consider. You will have to assess those risks in relation to their benefits.

DESIGN

Activity 1: Organize the Design Phase Work

Prior to initiating the actual work of the Design Phase, you should review your Study Phase effort and develop Design Phase details with all team members. Then you are in a position to organize the teams and assign them the actual tasks that are to be performed.

It is the role of the manager to see that the work flows well. For a large program, the manager will have a group of people in a systems office handling one or more project and support functions.

It is the role of the manager to see that the work flows well. The manager must be concerned with everyday details, including the following:

- Providing additional systems-related information where lacking
- Assisting in resolving system uncertainties
- Assisting in resolving new-member personnel conflicts
- Creating an atmosphere of open-mindedness

System-related information and uncertainties occur from the interaction of component parts of projects or programs of all sizes. For a small project, the manager will perform all the above functions. For a large program, the manager will have a group of people in a systems office handling one or more project and support functions.

Activity 1A: Review the Study Phase Effort

The Design Phase requires team members skilled in preparing documentation. Potential contractors are also added. Since the end of your Study Phase, time may have elapsed while the Design Phase funding is being approved. Review the Study Phase effort and its documentation in a project meeting with all involved persons.

Many of the added team members are responsible for preparing the design documents. They are often referred to as technicians, even though their skills may include design and development. These experienced individuals ensure that no technical or semilegal information is missing from documents. However, they are often accustomed to working with considerable guidance from the designers regarding design content.

PROJECT EXAMPLE

It is now the first Monday afternoon in January. Roger Stein, Jose Hernandez, and Professor Hulbert meet informally to discuss the situation of the additional $5 000 that is needed. Their results to date are not promising. They decide how Roger will present this information to the team the next day.

It is now the first Tuesday afternoon in January. Several potential new members have been invited to this first meeting of the Design Phase. Also invited by Professor Hulbert are two faculty members who have been consultants to local industry in the review of technical and semilegal documentation. Roger welcomes all returning team members back from vacation, and he introduces and welcomes the potential new members of the team. Most of the team members have already decided on their class schedules for the school term and share this information.

The team has heard rumors regarding the funding problem. Roger informs the team that a goal of $15 000 had better be set for the construction of the activity center. The additional $5 000 will have to include items that can be canceled without affecting the overall activity center construction and design costs.

Roger distributes the applicable Study Phase documentation to all students in attendance. He reminds them that their Design Phase details will be determined and assigned at their next meeting.

Activity 1B: Develop Design Phase Details

Develop the Design Phase content, where necessary, by performing the following:

PLANNING AHEAD ▶

- Preparing more detailed Design Phase task descriptions
- Devising a more extensive Design Phase schedule
- Allotting the Design Phase budget to individual groups

The groups are organized in activity 1C on the following page.

For most projects, delivery at or below the previously accepted cost is usually a requirement. However, some projects have deadlines that must be met. For these projects, the client may be willing to accept a cost greater than that originally contracted in order to meet a required schedule. Before spending additional money, obtain written agreement for any verbally discussed cost increase.

During the Design Phase, the design team must explore trade-offs between preplanned approaches and new ones now available. The Implementation Phase should proceed with a minimum (negligible) risk of significant change.

During the Design Phase, the design team must explore trade-offs between preplanned approaches and new ones now available. The Implementation Phase should proceed with a minimum (negligible) risk of significant change. Therefore, design trade-offs are examined to the level of detail required to minimize changes at a later stage. Design trade-offs must also be cost effective—the client wants the most "bang for the buck." Thus, all functions must be satisfactorily fulfilled at the least cost. Remember that sometimes an extra initial cost is acceptable to obtain a higher quality product that will have lower than expected maintenance costs.

Some tasks may require all the time allotted; other tasks can be performed faster. It may be possible to perform a task more quickly but at an additional cost. Thus, it becomes the responsibility of the project manager to decide the trade-offs among tasks, schedules, and budgets for the entire project.

Think about the nervousness that an astronaut must feel sitting in a complicated spacecraft on top of a rocket, knowing that all the craft's components were fabricated and assembled by the lowest bidders. The total system succeeds because the specifications, drawings, and plans have been carefully prepared by systems engineers who are responsible for devising a least-cost, high-quality, cost-effective system. You are learning to devise such designs via much smaller projects.

DESIGN

PROJECT EXAMPLE

Roger leads the team in devising a more detailed schedule for the next two months. The team members decide to assume an eight-week Design Phase because of the loss of days at the beginning of January as a result of the New Year holiday. The Design Phase tasks and budgets are also revised to match the detailed schedule. The new schedule is shown in figure 4.1.

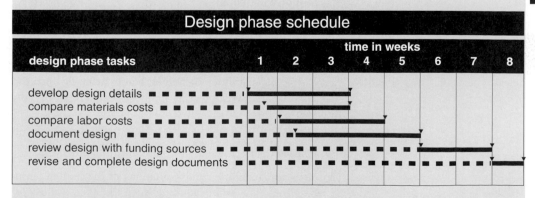

Figure 4.1 Design Phase schedule

Professor Hulbert and William D'Annolfo, Alice's uncle, meet and discuss how to guide the students during the Design Phase. They decide to invite two members of the school's Industrial Advisory Board who have been involved in many construction projects to become advisors and reviewers of the team's designs.

Roger and Jose agree with Professor Hulbert that they must focus on acquiring the additional $5 000. The team votes to keep Roger as team leader. For the Design Phase, he is considered to be the team manager. Therefore, he must also focus on seeing that the Design Phase schedule is met by all team members. Roger feels frustrated by this because he will not have much time to devote to the technical details. Professor Hulbert tells him that this is a common complaint of technical people when they are assigned the responsibilities of manager.

Activity 1C: Organize Teams and Assign Design Tasks

During the Design Phase, all of the Study Phase documentation is converted into the output documents that will guide the Implementation Phase.

Examine the skills, experience, and personalities of those who will be directly involved in the remaining activities of the Design Phase. Include in each group at least one person who has been involved in the Study Phase, to provide continuity from one phase to another.

Assign the various team members to developing each task, or group of tasks, for the Implementation Phase. This effort will convert all of the Study Phase documentation into the documents that will guide the Implementation Phase.

Develop the project plans for the selected project. Define the tasks in more detail so that each team member knows what is to be accomplished. Task descriptions are written and reviewed by the Design Phase group leaders and by those persons who will perform the tasks. Ideally, team members should have no questions about the following points relating to their assigned tasks:

- When a task should start
- What is to be accomplished by that task
- Which other tasks are to be coordinated with that particular task
- When that task is complete

Ensure that each group understands how its assigned design tasks interact with the design tasks of all other groups. The work breakdown structure and content of task descriptions are described in section 7.1 and appendix B.

Human factors (ergonomics) must be considered during the design of both equipment and software. The interactions of human operators or users with equipment and the environment are examined. These factors include the following:

- Location of displays and controls, meters, dials, and knobs on equipment
- Use of color coding
- Shape coding in both display icons and control knobs
- Work space design, including seating, location of work space, and illumination
- Effect of the environment on the users' performance and well-being
- Physical capabilities of human beings who will use the design; consider such factors as chair design, wrist position (to avoid carpal tunnel syndrome), and spinal column and back injuries
- Avoidance of distracting noises and colors where possible
- Expected sociological atmosphere and its effect upon operators

The physical capabilities and limitations of the human beings who will use your design must be recognized. The relationship between the effectiveness of operations of the system and the users' safety, comfort, and well-being must be kept in mind. See section 9.2 for further discussion of the influence of human factors on design.

It is the second Monday afternoon in January. Roger writes the six design tasks on the chalkboard. He asks the team members to write their names beside each of the tasks they prefer for their assignment. He revises the resulting list, with their agreement, as follows:

Design Phase Task	Assigned Personnel
Develop Design Details	Alice, Laura, Kim, Ellery, and Manuel
Compare Materials Costs	Laura, Jose, Manuel, and Joseph
Compare Labor Costs	Jose, Kim, and Joseph
Document Design	Alice, Laura, Kim, Ellery, Manuel, and Joseph
Review Design with Funding Sources	Roger, Jose, and Alice
Revise and Complete Design Documents	Alice, Ellery, Manuel, and Joseph

The team agrees to meet on Friday afternoons to evaluate status, problems, and plans. Laura agrees to document these meetings so any revised assignments are noted and equally distributed. Most of the team members agree to work on weekends when the CAD work stations at the university are more available. They will e-mail their progress to Laura on or before 8 A.M. on Monday mornings so that she can complete her memoranda for the Monday afternoon meetings. Thus, it is agreed that the team will also meet briefly on Monday afternoons to note any changes in status as a result of the weekend work.

Alice and Manuel interview Professor MacNevers, who teaches CAD. They discover that for construction of an activity center a series of drawings, known as a top-down breakdown, is needed. These drawings provide the required guidance for a contractor. Professor MacNevers notes that experienced contractors do not require a large amount of detail; however, students new to construction will need it (Shepherd, 1991). So it is decided that the students will prepare detailed drawings. The top-down breakdown list, with assignments, is as follows:

Plot Plan

    Landscaping (on a contour map):

        Alice and Laura

    Utilities Locations (water, sewer, gas, and electricity):

        Alice and Ellery

Building Plan

    Foundation Plan (with basement layout, including a natural gas burner location):

        Alice and Kim

    Framing Plan (with main floor and second floor—attic and ridge vent— including corner bracing and studding for doors and windows):

        Jose, Joseph, and Manuel

    Side Elevation:

        Manuel

DESIGN

Laura, Kim, and Ellery examine their class schedules and decide that Manuel will train them, as well as other interested team members, in the use of the CAD program. If they have trouble with learning the details, nonteam student assistants assigned to the CAD facility will help them in their use of the CAD program.

Remember: an important goal of the student project is to help the students gain project experience while providing more recreation for the children in the town of Bedford.

Remember: an important goal of this project is to help the students gain project experience while providing more recreation for the children in the town of Bedford.

Kim's Uncle Don, a retired human factors specialist, attends the second Monday afternoon meeting. He discusses how human factors considerations would apply to the activity center design (Sanders and McCormick, 1992; and Salvendy, 1997). They include the following from the ADA:

- Ramps for the handicapped and the range of slopes
- Space around toilets and sinks
- Colors that alert children to danger
- Location of doorknobs so that they are easily accessible to children
- Latches that are easy to use for doors children need to open
- Latches that are hard to use for doors that small children should not open
- Avoidance of sharp corners and edges that could injure the children
- Knobs that children with sight problems can distinguish by "feel"

Uncle Don also tells them of some of his experiences with human factors applications. The students invite him to attend their design review meeting, scheduled for the conclusion of the Design Phase.

The students decide to construct the foundation and first floor strong enough to support a usable second floor if the town ever decides to raise the roof and add a second floor.

Activity 2: Select the Solution Details to Be Implemented

A variety of implementation solutions should be examined before one solution is chosen. Vendors who will be providing materials must be contacted to determine the best schedule and cost options for their products. Make-or-buy decisions must first be identified and then one solution selected. Only after careful study of the entire design is it advisable to select one solution for the design.

Activity 2A: Study Potential Implementation Solutions

Studying implementation solutions involves analyzing required or desired functions and how they can be most efficiently implemented. The project is beginning to take shape via detailed drawings.

Examine each Study Phase document and then develop a design. Test this design against the Project Description, Design Specification, and Design Description that were prepared during the Study Phase. As each part in the design evolves, start to prepare the Production Specification. Test all recommended design changes against the original task descriptions, schedules, and cost estimates and against the expectations of your clients. Prepare a plan to develop the Working and Detail Drawings. Compare the effect of the recommended design changes and their details with your original task descriptions, schedules, and budgets from previous project phases (Nadler, 1970, p. 768).

Before proceeding to the design of a specific item or system, it is necessary to examine the availability of new materials, circuitry, and software that may not yet be described in even the most recent technical handbooks or literature. Changes occur so rapidly that it may be necessary to investigate "what's new" via the World Wide Web (Pugh, 1993, p. 177).

Begin the preparation of a schedule for assigning personnel and for delivery of materials for the Implementation Phase. The schedule should include the following:

- Project personnel assignments
- Contractors' performance
- Materials deliveries
- Vendor deliveries

It is this second and more detailed schedule that may cause changes in the preliminary schedules. Why? Because people (including contractors) may not have the staff or facilities to work on more than one project at the same time. Merge all schedules into one master schedule before you complete the Design Phase.

For the Implementation Phase, the following items should be analyzed:

- Estimates of project staffing and additional personnel needed
- Estimates of work space needs
- Required project furnishings, fixtures, and equipment
- Needed project-support supplies and materials, such as portable toilets

Indicate these items in a separate document rather than as a portion of the actual product implementation documents. They are considered as a part of your project overhead. As such, the costs are not presented to the client.

PROJECT EXAMPLE

Professor Hulbert reminds the team members that they must examine each Study Phase document and recommend any needed changes.

- Alice reviews the Project Description. She recommends some changes relating to activity center costs and schedules.
- Laura reviews the Design Specification. She recommends a few changes regarding the concrete foundation design.
- Roger reviews the Design Description. He recommends minor changes regarding the worker skills required to assist the team.

Three potential design directions are considered. Each design direction is tested against the revised Project Description, Design Specification, and Design Description documents prepared as a part of the Study Phase effort.

Roger initiates early preparation of the Production Specification as each design form evolves from its predetermined function.

Here is an example of how "form follows function" in the activity center project:

Function: Children require "busy" activities to learn; amusement is intertwined with learning.

Form: An activity center provides a physical structure where children can meet, learn, and "have fun" while learning.

Thus the function of the activity center is established, and the form of the solution then follows.

DESIGN

The exact building materials to be used are examined. The following questions are asked:

What functions (uses) can be accomplished by the various designs, and which functions must be revised, deleted, or added?

Is the total cost of the project still between $15 000 and $20 000? (Additional money may not be obtainable from the funding sources.)

What type of foundation will be required? Will it contain steel reinforcing rods and less concrete, or only concrete?

What is the type of soil and/or ledge, and how will it affect design and costs? (Several holes must be dug, and earth extracted and tested. Local contractors may have knowledge of the soil conditions.)

Could a second floor be added later? (This is considered because the town has recently noted a possible future need.)

Will the partitions be fixed or movable?

Roger reminds all the team members that the project may not be funded if the costs become excessive.

Roger assigns Jose, Kim, and Manuel the extra responsibility for monitoring the changes in materials costs and labor costs by applying Manuel's newly devised computer program. Manuel developed it for a software class and designed it to meet the needs of the activity center project.

Each design change under consideration is tested weekly. The test is against these factors:

- The original task descriptions
- Schedules
- Cost estimates
- The expectations of the town of Bedford representatives
- The expectations of Miss Kewer's attorney

The team members are told they must assess the effect of their recommended design changes against the above list. Their recommended changes will be considered during the weekly Monday afternoon meetings. They are told to provide them to Laura via e-mail on or before 8 A.M. Monday. This time allows her to prepare a hard copy for each person attending the afternoon meeting.

Ellery agrees to begin preparation of a Personnel Assignment and Materials Delivery Schedule for the activity center Implementation Phase. This more detailed schedule is to include the following:

- Project personnel assignment dates
- Contractor performance dates
- Material delivery dates
- Vendor delivery dates

Ellery also agrees to provide estimates for the following:

- Project staffing and additional personnel needed
- Work space needs during construction
- Project furnishings, fixtures, and equipment
- Project-support supplies and materials (such as portable toilets)

Alice and Joseph agree to review Ellery's effort each Monday at noon.

One area that must be investigated is how to connect the carrying beams to the exterior wall studs. One suggestion is that metal plates be used. Alice and Joseph have some background knowledge about this construction issue. They agree to perform the following tasks:

- Determine sources for metal plates
- Investigate availability of metal-plate length, width, and thickness
- Determine the bolt size and type needed to fasten the various metal plates
- Estimate the time required to purchase and install metal plates
- Estimate the labor required to install the metal plates with their associated bolts

Manuel agrees to review their effort every other Monday at noon.

Manuel also agrees to investigate the following:

- Types of plumbing materials available
- Types, brands, and cost of paint

Joseph agrees to review his effort every other Monday at noon.

Ellery is assigned to document all changes to the Implementation Phase schedule. Alice and Joseph are assigned to review the vendor data and recommend which vendors will be used for the Implementation Phase.
Roger offers to outline the contents of the Working and Detail Drawings.

Activity 2B: Contact Vendors for Make-or-Buy Information

Explain to your potential vendors, by means of both meetings and documents, those areas where you may purchase an item from them rather than fabricate it yourselves. Obtain vendor input regarding costs and delivery schedules for each item you are considering for a make-or-buy decision.

DESIGN

PROJECT EXAMPLE

At this point in the project development, vendors can be contacted for bids and the extent of work they will perform can be determined.

Alice and Joseph receive copies of the drawings in their draft form from team members who are preparing them. The pair will deliver the drawings to the vendors for use when they prepare their initial bids for the project. Vendors' reputations are examined through the local Better Business Bureau, and the team assesses the vendors for potential use as part of the final selection process within the Design Phase. Potential vendors are informed that they will be selected or rejected on or before March 10. Information about each vendor should include the following points:

- Quality of work
- Dependability regarding schedule
- Excuses offered that increase costs
- Willingness to adapt to schedule and design changes that may occur

The vendors are requested to respond within two weeks and note on their return bids/quotes: "These prices are firm for 60 days."

Figure 4.2   Make versus buy comparison: set of stairs

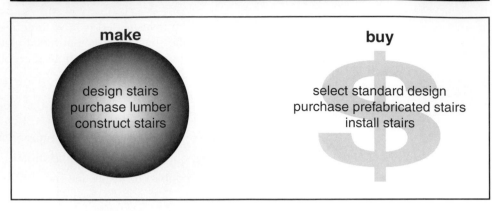

Figure 4.3   Make-or-buy sequence

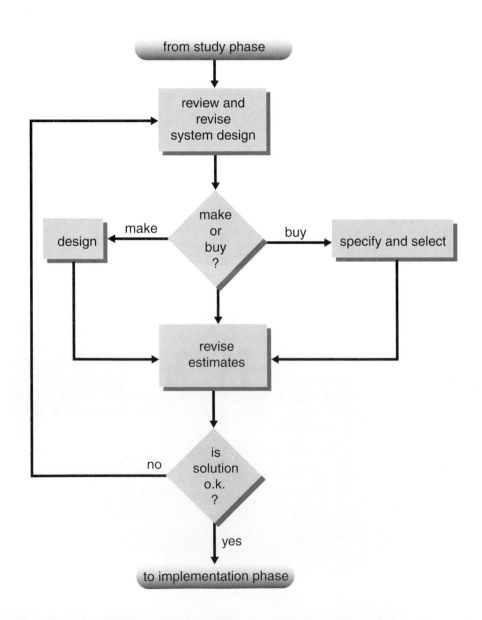

Activity 2C: Determine Make-or-Buy Decisions

Examine all the make-or-buy information you have gathered. With this information, you can now do the following:

- Develop design details for the project
- Review and revise the overall design where necessary
- Decide what portions will be fabricated by the designers ("Make")
- Decide what portions will be purchased from vendors ("Buy")

Revise the content of the task descriptions, as well as estimates of costs and schedules, to agree with your make-or-buy decisions.

A "make-or-buy" decision concerning the construction of a set of stairs can serve as an example. (See figure 4.2.)

Make: Students design the stairs, purchase the lumber, and construct the stairs.

Buy: Students (school) purchase prefabricated set of stairs from a lumber supplier.

All the following factors must be considered: total labor and materials costs, convenience, and availability. If the cost differential is close between using the original contractor versus buying an item from another source or performing the work oneself, it may not be worth the effort to "go outside" to a contractor. Also be careful not to invest more time (and, therefore, money) in reaching a decision than will be saved by awarding the work to a contractor.

The make-or-buy decision sequence is given in figure 4.3.

PROJECT EXAMPLE

Laura and Kim devise a matrix of information comparison for the activity center trade-off decisions. For the activity center, the areas for decision making are shown in figure 4.4.

Information comparison matrix

item	initial costs labor material	annual cost of maintenance	life expectancy	exterior appearance
framing lumber:				
pine				
white cedar				
concrete blocks				
poured concrete				
roof shingles				

Figure 4.4   Information comparison matrix

DESIGN

- Which will cost less—lumber, concrete block, or poured concrete—for the foundation? (Labor and material costs, as well as appearance, must be considered.)
- Where is the best trade-off regarding the quality of roof shingles? (Cost versus life expectancy is primarily a quality judgment.)
- What is the preferred exterior finish: clapboards or wood—such as white cedar—shingles? (Consider initial cost, maintenance costs, and desired appearance.)

These and other design trade-offs are examined to the level of detail required to ensure that implementation can proceed with little risk of significant change.

Laura and Kim distribute an "Activity Center Make or Buy" document that they have developed. It is reviewed during a Friday afternoon meeting. The decisions are documented and the results are distributed. Then the appropriate vendors are informed.

At this point in the project, design trade-offs are examined to the level of detail required to ensure that implementation can proceed with little risk of significant change.

Activity 3: Document the Design Solution

It is time to prepare the Production Specification, as well as the Working and Detail Drawings. In effect, the project design and the reason for its choices are being documented.

It is time to prepare the Production Specification, as well as the Working and Detail Drawings. Refine the forthcoming Implementation Phase tasks, schedules, and budgets. (See chapter 7.) In effect, you are documenting your project design and the reason for its choices.

Activity 3A: Prepare the Production Specification

The Production Specification originates from the Design Specification, prepared during the Study Phase. Here are suggested steps for preparing this document:

1. Study the Design Specification.
2. Devise, as necessary, an outline of the Production Specification.
3. Extract, wherever possible, the appropriate portions of the referenced specifications for inclusion. The Design Specification may only refer to environmental and other more general specifications, rather than contain the actual wording.
4. Modify the Design Specification paragraphs to become new paragraphs within the Production Specification.
5. Refer to the Working and Detail Drawings where appropriate.
6. Review and revise the results, where necessary, during activity 3D.

The Production Specification may include the applicable excerpts from the referenced general specifications. Typical specifications contain the following paragraphs or groups of paragraphs:

- Scope of Specifications
- Applicable Documents
- Requirements
- Quality Assurance Provisions
- Preparation for Delivery
- Notes and Appendixes

Government specifications are the most complex; industrial specifications are usually much simpler. See section 8.1, page 205.

Specifications, including those for construction, follow standard formats. They ensure that contractors perform their portions of the project according to what the client wants.

Alice and Ellery convert the Design Specification into the Production Specification. The content evolves, with paragraph headings added, as noted below. Refer to the specifications for the activity center given in appendix C. Each section consists of paragraphs identified by letter. Subparagraphs are identified by number.

The following sections in specifications are often referred to as nontechnical boilerplate. They are standard in project documents.

General Specifications contain the nontechnical definitions required to avoid misunderstandings later. This section also identifies the organizations and persons who are specifically responsible for the performance and monitoring of the project.

General Requirements contain nontechnical information on who will do what. This section also defines the legal responsibilities, as required by the appropriate local, state, and federal agencies. Workmanship is described, along with how the selection of "equivalent" materials or equipment will be approved.

Intent of Contract Documents is included. Thus if there are errors or duplications that have been overlooked during the other phases of the project, they are to be immediately brought to the attention of all parties involved. At that time, the issues can be promptly resolved.

Subcontractors should not become the problem of the client. This section specifically states that the (prime) contractor is totally responsible for the actions of any and all hired individuals and organizations.

Project Meetings are the responsibility of the contractor. Thus, the contractor assumes full responsibility for coordinating all efforts to ensure that work is performed correctly and on time.

Cleaning requires a separate section because many contractors will leave a mess for others to remove unless they are told otherwise.

Workers' Compensation and Insurance has become important because workers may become injured while working on the project. Legal action in the form of lawsuits can be filed against all parties involved with a project. Contractors must purchase liability insurance for their protection.

Project Closeout must also be defined, and proof given to the client that there are no outstanding debt claims that could be filed against the client. Operating and maintenance manuals and warranties must be delivered to the client so the fixtures and equipment can be properly operated and repaired.

The following sections form the technical content of specifications.

General Site Work is the beginning of the technical portion of the specification. This section is an overview of the remaining technical sections.

Earthwork describes the preparation of the site prior to construction of the foundation and includes specifications for later grading, drainage, and soil-and-rock storage and removal.

Landscaping describes how the earthwork around the buildings shall be finished, graded, and seeded.

The **Site Utilities** section defines the responsibilities of the contractor regarding installation of sewer, water, natural gas, electricity, and telephone service.

Concrete Formwork is an overview of the next three sections. It also notes that forms shall be properly installed and braced, which is a safety feature.

DESIGN

Cast-in-Place Concrete describes the exact sizes, strengths, and consistencies of the footings, walls, and floors, and what to do when ledge is encountered.

Dampproofing is important because many structures are being constructed in areas where water is either present or can accumulate during and after rainstorms or snowstorms.

Subsurface Drainage System requirements exist to remove any water that may accumulate around and under the foundation.

The **Framing and Carpentry** section contains all the wording required to constrain the contractor in the construction of floors, ceilings, roof, and walls. It also, where appropriate, suggests preferred types and sources of material, such as grades of lumber. It also refers to the accompanying drawings.

The **Roof** section describes the types of roof materials to be used and how they are to be installed.

Window and Door Schedule lists the windows and doors to be purchased and how they are to be installed and sealed.

Exterior Trim and Siding defines the type of exterior trim and siding to be installed and the sealing material to be applied.

Plumbing and Heating describes the materials and procedures for water and sewer installation; installation of bathroom and kitchenette fixtures; equipment, materials, and procedures for installation of a furnace and ducts or pipes; the type of central heating, or other form of heating, ventilation, and/or air conditioning (HVAC), as well as related equipment and its installation, to be used to control the environment.

The **Electrical** section lists the equipment and service to be provided via an underground service pipe.

The **Walls, Floors, and Ceilings** section defines the materials to be used on the bathroom and other walls, floors, and ceilings.

Locks are described in detail. (For the activity center, for example, provision is made so access may be controlled by a town employee through a master key to a deadbolt lock.)

Landscaping, Driveway, Parking, Ramp, and Walks are carefully specified in order to provide for vehicle access and parking, walkways, and removal of water.

There are standard specifications and contract forms for construction projects, which may be purchased from the AIA (The American Institute of Architects, 1735 New York Avenue NW, Washington, D.C. 20006) at www.aia.org or the AGCA (Associated General Contractors of America, 333 John Carlyle Street, Suite 200, Alexandria, VA 22314) at www.agc.org.

Activity 3B: Prepare the Working and Detail Drawings

Study your Design Description. Devise, as necessary, a list of drawings to be prepared, typically entitled "Top-Down Breakdown." (The development of this list of drawings is usually iterative.) Expand your preliminary sketches. See figures 4.8 and 4.9 on pages 112 and 113. Provide details for each of the portions of your project, as well as interface information between and among interconnecting and interrelated items. Refer to the Production Specification where appropriate. Review and revise your results, as necessary, during activity 3C on the next page.

The Working and Detail Drawings originate with the Design Description in the Study Phase and its accompanying preliminary sketches. The Design Description should contain general information. The Working and Detail Drawings include specific information as needed.

PROJECT EXAMPLE

The Working and Detail Drawings, as assigned in the Project Example for activity 1C, should use a set of notation standards so all team members use the same layout techniques and notations. Professor Hulbert recommends that decisions regarding format be documented and distributed so all team members will be able to understand the individual drawings. The following drawings are prepared and reviewed:

Plot Plan (a perspective view that defines the boundaries of a parcel of land and the location and orientation of the structure to be built)

Landscaping

Location of Utilities (water, sewer, gas, and electricity)

Building Plan

Foundation Plan (including basement layout with furnace location)

Framing Plan (with main floor and potential second floor—attic and ridge vent—including corner bracing and studding for doors and windows)

End and Side View

The above drawings are reviewed by Professor Hulbert, all team members, and Professor Jacobs (who is also an Architectural Engineer). There is now enough documentation to allow the preferred vendors to prepare final bids for construction of portions of the activity center.

Activity 3C: Refine Tasks, Schedules, and Budgets

The final documentation for the Design Phase must include revised and detailed descriptions of tasks, schedules, and budgets.

Review the Implementation Phase task descriptions with respect to any changes that result from your review with the funding source. Revise the task descriptions where necessary. Verify that the individuals and groups who will be responsible for the Implementation Phase agree on the meaning of their assigned task descriptions. (See sections 7.1, 7.2, and appendix B.)

Prepare your final Implementation Phase tasks, schedules, and budgets after questioning those persons who will be responsible for each task. Verify or revise total costs after examining the new budgets. Review how tasks and schedules intertwine. Revise the tasks and schedules as required. Then negotiate any necessary task and schedule changes with those team members or contractors who will be responsible for them.

Review all costs related to the completed design. Verify that the costs are acceptable to the team members or contractors to whom each task will be assigned. Only then is it possible for you to be certain that no one can later say, "I didn't know that" or "It cannot be done for that amount of money."

PROJECT EXAMPLE

The work of the design for the activity center as reviewed is considered complete. During the latter part of the seventh week and the entire eighth week of the Design Phase, the final documentation must be prepared. As the team members review the results of the previous activities, they discover, to their embarrassment, that an extra two weeks of time will be needed to construct the activity center, and the project will extend into September. The team members

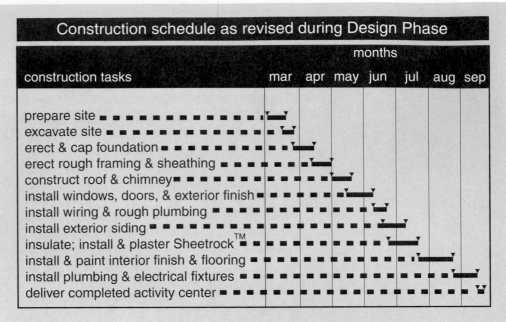

Figure 4.5   Construction schedule as revised during Design Phase

have contacted many vendors of materials and contractors of specialty areas. The new schedule for the Implementation Phase is shown in figure 4.5.

Note that the titles of some of the tasks have been revised slightly. Thus, the titles of the Implementation (Construction) Phase cost estimates also have to be revised so they agree with the new schedule titles. The cost estimates must also be changed so these estimates indicate when the contractors are to be paid, as shown in figure 4.6. Contractors are normally paid at the end of each major task or at the end of a month.

Revised summary estimate for the Implementation (Construction) Phase

construction tasks	months							
	mar	apr	may	jun	jul	aug	sep	totals
prepare site	600							600
excavate site	1400							1400
erect & cap foundation		1700						1700
erect rough framing & sheathing		600						600
construct roof & chimney			5100					5100
install windows, doors, & exterior finish				1500				1500
install wiring & rough plumbing				1400				1400
install exterior siding					1600			1600
insulate; install & plaster Sheetrock™					1300			1300
install & paint interior finish & flooring						2000		2000
install plumbing & electrical fixtures							1800	1800
deliver completed activity center							1000	1000
task totals	2000	2300	5100	2900	2900	2000	2800	20 000

Figure 4.6   Revised summary estimate for the Implementation (Construction) Phase

Note: Jim Sullivan (plumber) and Jim Stander (electrician) offer to donate some of their time to keep the total costs within the $20 000 allotted for the project.

Had two contractors not offered to donate their services, the total Implementation Phase costs would have been greater than $20 000. In order to avoid losing the project, two contractors who live in Bedford—one skilled in plumbing (Jim Sullivan) and the other in electrical work (Jim Stander)—offer to donate their services so that the children of the town will have an activity center. (Stander has agreed to perform all wiring and fixture installation, using the labor of various team members.) After discussing the offers with the appropriate town officials, the team accepts the offers of donated labor. A note on the proposals indicates that contribution as a part of the contract.

A decision is made, as advised by Professor Hulbert, to include the actual cost of donated labor as a zero-cost item in those estimates where labor is a separate subtask item (not shown). Thus, for student learning purposes, a more realistic cost estimate of the total project exists.

Activity 3D: Document the Design and Reasons for Its Choice

The completed documents from the Design Phase are sent to the contractors who will use them during the Implementation Phase. These consist of a Letter of Transmittal, the Production Specification, and Working and Detail Drawings.

The completed documents from the Design Phase consist of the following:

- Letter of Transmittal, which for the activity center will be a Contract Letter
- Production Specification
- Working and Detail Drawings

The content of each of these documents is described in section 4.3.

Prepare the Letter of Transmittal (or Contract Letter). The contents of this document are described in section 4.3. Before this document is signed, it should be reviewed by the client, the team manager, and by appropriate legal personnel.

Complete the Production Specification. Contents for this document are described in detail in section 8.1 on page 205.

Devise either a bill of materials or a parts list, and prepare the Working and Detail Drawings. The Working and Detail Drawings consist of the following:

- Presentation Drawing, containing little or no structural information
- Models where appropriate
- Working Drawings, containing all necessary details

Depending on the product or software to be delivered, the drawings may require several sheets. See the next section for more information. When only experienced contractors are allowed to bid on a project, the need for detailed information may be greatly reduced.

As noted in chapters 5 and 7, documents must be prepared at the end of every phase. All *deliverables* must be defined and prepared. What are deliverables? According to Duncan (1996) a "deliverable" is a tangible, verifiable product such as a feasibility study, a detailed design, or a working prototype. The conclusion of a project phase is generally marked by a review of deliverables (such as end-of-phase documents) and project status. Thus, it is possible to (a) determine if the project should continue into its next phase and (b) detect and correct errors in a cost-effective manner. *Deliverables* include both equipment and documents that are *delivered* to the client.

Deliverables that may not be quite so obvious include copies of government-required permits for wiring and plumbing, equipment and its operation and maintenance manuals, recycling data to be implemented at the end of equipment life, and copies of warranties for equipment purchased by the contractor. These items are installed as part of the project implementation; their warranties are passed along to the client to protect the client in the case of a failure.

DESIGN

Costs for the Implementation Phase are now known quite accurately. Also, the money that was spent during the first three phases is known. Now Marketing can establish a price for any product that will be produced in quantity and determine if the product will sell at that price and provide a profit.

PROJECT EXAMPLE

Laura contacts a professor at the Northeastern School of Law to assist in writing the Contract Letter. See appendix D for a copy of the final version of this document. The letter is reviewed by each team member before it is shown to the town of Bedford selectmen. The legal wording is new to most of the team members, so the students discover that reviewing it is a good learning activity.

Alice and Ellery prepare copies of the Production Specification that they have been developing as described in activity 3A. The document's contents are given in appendix C. The copies are distributed to all members of the team, the law professor, and all involved contractors for review. Alice and Ellery revise the specifications as required.

Roger decides to review and coordinate the final Working and Detail Drawings so he can feel he has participated in some of the technical effort for this phase. A computer-generated plan view drawing of the activity center, without dimensions, is given in figure 4.7.

All three sets of documents are gathered together, reviewed by the team, and prepared for distribution to contractors. A letter that explains the contents of the package is included. The letter also reminds all involved parties that the schedule must be followed and the costs kept within budget. It also indicates that Roger, Jose, or Professor Hulbert is to be contacted immediately if any questions arise.

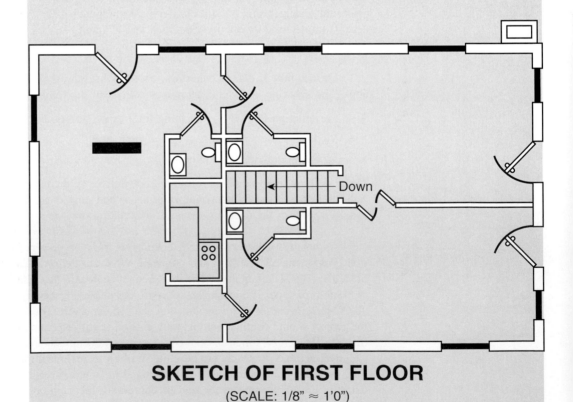

SKETCH OF FIRST FLOOR
(SCALE: 1/8" ≈ 1'0")

Figure 4.7   Plan views of activity center

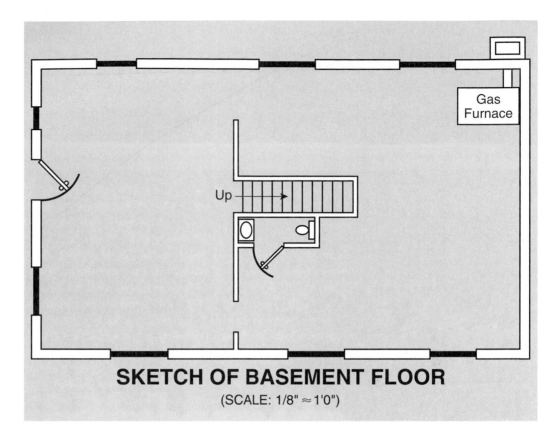

SKETCH OF BASEMENT FLOOR
(SCALE: 1/8" ≈ 1'0")

PLANNING AHEAD▶

Activity 4: Prepare for the Implementation Phase Work

It is time to review the Design Phase results. You should also verify the availability of Implementation Phase funding and then complete the planning for the Implementation Phase.

Activity 4A: Review the Design Phase Results

As the design is being developed, it needs to be reviewed by experts and clients for revisions and oversights.

The Design Phase results must be carefully reviewed and compared with the Study Phase documents. Here is one sequence for accomplishing this task:

1. Examine your newly designed product or software from a total system viewpoint.
2. Compare this design with the Study Phase documents: Project Description, Design Specification, and Design Description.
3. Verify that the Design Phase forms match the Study Phase functions.
4. Verify that the ground rules (Study Phase activity 2) were followed.
5. Review any revised ground rules.

DESIGN

PROJECT EXAMPLE

Roger requests a formal review meeting with Professor Hulbert and other faculty members chosen by the professor. These faculty members are given a copy of the design ground rules so they can compare the results to the original requirements. Two administrators from towns adjacent to Bedford are invited to attend the review because of their experiences with similar structures for children.

The meeting occurs at the end of the fourth week of the Design Phase so that any recommended changes can be included prior to presentation to the funding sources. Only five significant changes are recommended during the four-hour review meeting. They are included as design changes during the fifth week.

Activity 4B: Verify Availability of Implementation Phase Funds

The detailed design and plans need to be presented to the funding sources for approval and review.

Inform funding sources regarding the status of the Design Phase. Prepare and give a presentation if necessary. Obtain both general and detailed reactions from the representatives of the funding source. Verify the availability of the needed Implementation Phase funds and their payment schedule.

PROJECT EXAMPLE

Roger reports that the existing $15 000 funding has remained firm. He further notes that an additional $2 000 has been pledged by last year's Northeastern team, which had a surplus of funds from their project. An additional $1 000 is pledged by five faculty members at $200 per faculty member. Roger reminds the team that $2 000 more needs to be raised for the desired $20 000 required to implement the existing plans. He also stresses to team members that they must continue to keep costs under careful control.

At the beginning of the sixth week of the Design Phase, Roger gives presentations to those groups and individuals who will be responsible for the funding of the Implementation Phase. A briefing is scheduled for the participants for Wednesday evening at the university. At that time, Roger describes the drawings for the activity center, shows a scale model of the center, gives a summary of the center specifications, and distributes the Contract Letter for review. He asks the attendees to determine if anything has been overlooked. He also presents the team's assumptions regarding the personnel who will be occupying and working at the activity center. He notes the other items that still need to be prepared include the following:

- An estimate of staffing and additional personnel needed to complete the project

- An estimate of work space needs for the activity center

- A description of preferred furnishings, fixtures, and equipment

- An estimate of needed toilet facilities and storage space, such as bathrooms for adults and children, shelving for worker use, lockable closets and cabinets.

The persons involved in the funding agree, in advance, that the design looks acceptable. (Thus, there is less of a chance that changes will be requested once construction is in process.) The meeting ends with all participants eager to see construction start, and they offer to assist in locating the additional $2 000 needed.

Activity 4C: Plan the Implementation Phase

Before the Implementation Phase begins, the plans should be reviewed to see that they are realistic. All participants should realize that design work stops once the Implementation Phase begins.

Verify that the sequence of work to be accomplished during the Implementation Phase is realistic. Devise and verify the content of a schedule of personnel assignments and material delivery that coincides with the Implementation Phase tasks and schedules. Management subtasks (Implementation Phase activity 3A) should be planned at this time.

Once the Implementation (Construction) Phase starts, any revisions to previous phases will be both expensive and time-consuming. You should make every effort to have all your designs and documentation completed during the Design Phase. The team should not continue designing once the Implementation Phase is initiated.

There is an expression often used by engineers and technologists: "Never deliver drawings and specifications to a manufacturing organization and then go away and forget them." Personnel in manufacturing may need assistance to interpret the words and drawings developed by the design team. Why? Because English is *not* a precise language for communication purposes. Thus, some design-team members and project management personnel need to be a part of the Implementation Phase to ensure project continuity.

It is the eighth week of the Design Phase. As the final documents are being prepared and reviewed, Roger, Jose, and Professor Hulbert decide to offer their "secret" plan to the team for raising the remaining $2 000. They are going to have a weekend open house in Bedford at the high school gymnasium and request a minimum donation of $5 from individuals and $10 from families to see the plans and model on display. The Bedford Farms Dairy has agreed to donate soda and ice cream for the open house. The local Bedford newspaper provides free advertising for the event.

At the open house, the construction plan is presented once every half hour by Roger and Alice. In addition, Manuel gives a continuing computer demonstration so that the children and parents may see how changes to the activity center design can be quickly documented. His computer is connected via a modem to a Northeastern University work station. (Jose assisted in the development of Manuel's demonstration.) As a result of the event, an additional $3 500 is raised. The team and Professor Hulbert are delighted.

The final assignments for the Implementation Phase can now be scheduled. All funding sources, town selectmen, vendors, Alice's uncle, supporting faculty, and the president of Northeastern are informed of the fund-raising success.

The Personnel Assignment and Materials Delivery Schedule for the Implementation Phase are prepared by Ellery Kistler. They include the following:

- Individual work assignments and schedules for team members
- Contractors' schedules
- Materials delivery schedules
- Vendor delivery schedules

It is this more detailed schedule that may cause changes in the more general Implementation Phase schedule. Vendors, including contractors, may not have the staff to work on two or more projects at the same time. Alice contacts the selected vendors and contractors to ensure that they will be available as needed according to the now final Implementation Phase schedule.

The plans for the Implementation Phase can now take the form of numbered tasks and subtasks to which precise budgets and schedules are assigned. The documentation for the project is now the most complete and accurate.

Ellery and Alice decide that the tasks and subtasks are now specific enough to be numbered. As a result, Manuel can easily connect them to the budget and schedule. Together the three prepare and publish the following information in the form of a schedule:

1. Prepare Site
    1.1   Grade a rough roadway to the activity center site
    1.2   Clear and remove excess soil, trees, and brush as necessary
    1.3   Establish a water supply and portable toilets for workers
    1.4   Provide temporary electrical power for the construction equipment
    1.5   Locate municipal sewage line or sewage-disposal tank
    1.6   Survey and install stakes to identify location of the activity center foundation
2. Excavate Site
    2.1   Excavate for footing, foundation, and chimney base
    2.2   Survey and install batter boards with reference nails and outline twine
    2.3   Establish top level of footings; excavate for spread footings

DESIGN

3. Erect and Cap Foundation; Install Sills
   3.1   Install spread-footing forms
         Pour spread footings
         Remove forms for spread footing
         Install foundation forms and reinforcing bars
         Pour foundation walls
         Remove forms from foundation walls
   3.2   Install wooden foundation sills

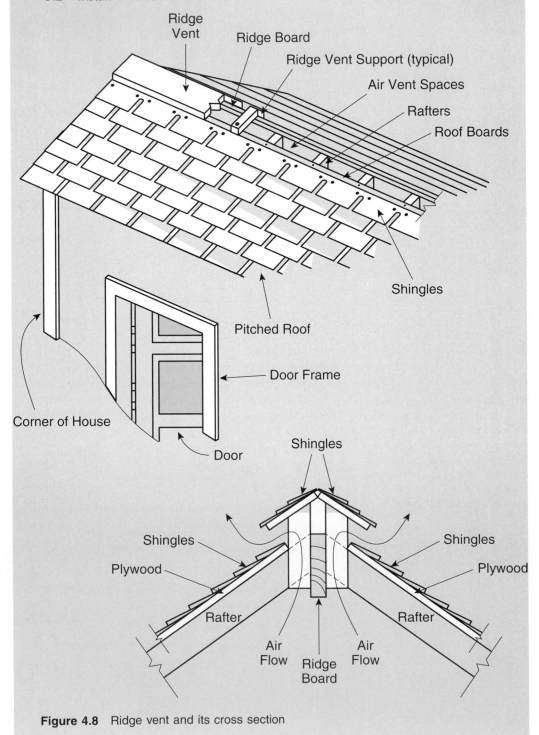

Figure 4.8   Ridge vent and its cross section

4. Erect Rough Framing and Sheathing

    4.1   Install floor joists and bridging

    4.2   Install rough flooring on floor joists

    4.3   Erect exterior side walls (including studs with openings for windows and doors)

    4.4   Erect interior walls (with openings for interior doors) and rough stairs

    4.5   Install wall caps

    4.6   Install ceiling joists

    4.7   Install exterior wall material (sheathing on studs)

5. Construct Roof and Chimney (See figures 4.8 and 4.9)

    5.1   Erect roof rafters

    5.2   Erect end wall studs to roof rafters; sheath ends of roof

    5.3   Close in the roof and install vent-pipe flashing

    5.4   Construct chimney with roof flashing

    5.5   Install shingles on the roof

6. Install Windows, Doors, and Exterior Finish

    6.1   Install prehung windows and flashing

    6.2   Install prehung exterior doors and flashing

    6.3   Install finish molding or boards at corners and roof edge

    6.4   Install soffits and fascia

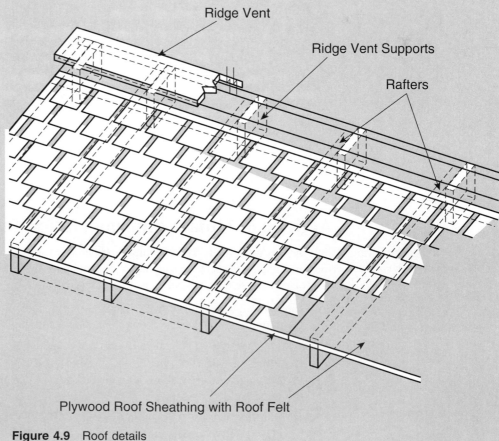

Figure 4.9   Roof details

DESIGN

7. Install Wiring and Rough Plumbing
   7.1 Install external kilowatt-hour-meter base
   7.2 Install interior circuit-breaker panel
   7.3 Install internal wiring, including boxes for outlets, switches, and lighting fixtures
   7.4 Install rough plumbing and plumbing vent pipes
8. Install Exterior Siding
   8.1 Install exterior finish on walls
   8.2 Prime and paint exterior finish, walls, doors, and windows
9. Insulate; Install and Plaster Sheetrock™
   9.1 Install interior insulation between studs of exterior walls
   9.2 Install insulation and strapping in ceiling
   9.3 Install and plaster ceiling Sheetrock™
   9.4 Install and plaster wall Sheetrock™
10. Install and Paint Interior Finish and Flooring
    10.1 Install finish on exterior door and window frames
    10.2 Install baseboards
    10.3 Install finish flooring
    10.4 Sand, seal, and paint floors
    10.5 Hang prehung interior doors
    10.6 Paint interior ceilings, walls, and finish
11. Install Plumbing and Electrical Fixtures
    11.1 Install plumbing fixtures
    11.2 Install electrical fixtures, including switches and outlets
    11.3 Install gas heater

Manuel modifies a computer program related to construction to prepare the project documentation. He adds columns that include the names of individuals, vendors, and contractors with their delivery dates.

Activity 5: Obtain Implementation Phase Funding

All the steps necessary to obtain funding and to obtain formal agreement to the plans from funding sources and clients must be completed at this time in the project.

At this time, the tasks, schedules, and budgets for a project will be the most accurate. Any further changes will occur during the Implementation Phase. Now it is imperative to obtain funding for that phase. The typical steps to obtain that funding are the following:

1. Review refinement of tasks, schedules, and budgets; design documentation; and Implementation Phase plans with your client and/or funding source. See Design Phase activities 3C, 3D, and 4C.
2. Agree on the content of your Letter of Transmittal or Contract Letter.
3. Obtain signatures from those individuals who are designated to be legally responsible for the work and its payment.
4. Review your construction and/or delivery and payment schedule as applicable.
5. Prepare to initiate the Implementation Phase.

Terminate the Design Phase effort as soon as the work has been completed and reviewed. A purchase order or other authorization should have arrived for funding the Implementation Phase. If a foundation is a source of funds, then a check for the Implementation Phase will be provided at this time.

PROJECT EXAMPLE

A Letter of Transmittal, including the Letter Contract previously reviewed (see activity 3D), is hand-carried to each authorized person within the state and town and to Eleanor Kewer's trust-fund attorney. Once all signatures are obtained, signed copies of the contract are mailed to each funding participant.

A charge number for expenditures is opened by Professor Hulbert with the university treasurer. A workplace trailer-office and a portable toilet are located. They will be delivered to the Bedford construction site on March 1. A bank account is opened via the university so the university treasurer can verify all invoices and authorize their payment.

It is the end of February. Roger terminates the Design Phase. He sends letters of thanks to all team members, to all those who donated to the project Implementation Phase, to the Bedford Farms Dairy, and to the local newspaper for publicizing the team's open house.

SECTION 4.3   DESIGN PHASE DOCUMENTS

The documentation prepared during the Design Phase will be used to guide the Implementation Phase. Typically the Design Phase documentation is delivered from the engineering team to the manufacturing team. As noted in figure 4.10, manufacturing personnel may be assigned to the Design Phase team to participate in the development of documentation. Since humans read and interpret documents, documents run the risk of being misinterpreted. When the document users are involved in document preparation, this helps to reduce the chances of misinterpretation. Likewise, engineering personnel are assigned to the Implementation Phase manufacturing team to assist in documentation interpretation.

There are at least two approaches to ensuring a quality product. One approach is to invite some of those persons who will be involved in the manufacturing process to become a part of the design team. Another approach is to designate in the early phases of the project those individuals who are involved in the design phase as persons who will become a part of the manufacturing team. Include persons familiar with the various aspects of maintenance necessary for support of the final product. This may include **out-year** requirements (after warranty expiration). Illustration: Many a kitchen in a slab-foundation ranch-style home has been ruined by leaking pipe joints buried in concrete. No one planned for out-of-warranty access to these potential failures.

DESIGN

Figure 4.10 Interaction between the design and manufacturing teams in preparing documentation

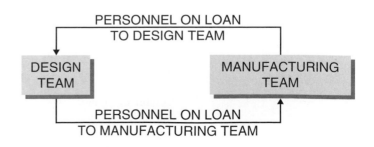

Figure 4.11 Maintenance factors

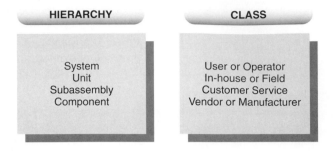

HIERARCHY	CLASS
System Unit Subassembly Component	User or Operator In-house or Field Customer Service Vendor or Manufacturer

Consider both the hierarchy (level) and class (type) of maintenance needed (figure 4.11). Audio units, such as CD players, mixers, speakers, and receivers connected together become a specific audio system. Unit vendor warranty is of no use until system maintenance determines a malfunctioning unit. On-scene, or vendor, maintenance personnel must test internal circuitry to isolate unit failures. Circuitry may be located on replacement type subassemblies or hardwired motherboards. Maintenance time is reduced when unit developers plan for ease of access. Identified faulty subassemblies or individual components are then replaced. Orientation, physical size, and electronic specifications must be duplicated. Good maintenance planning ensures these parts are available for a specific time frame.

Illustration: After a motor burned out in an HVAC system for the third time, customer service representatives began their own research. Nowhere in the prototype-system test documentation was this defect recorded. Neither was it recorded as a problem in any literature. They discovered that the documentation delivered to the client did not contain any user, or operator, requirements to lubricate this motor's unsealed bearing every ninety days (normally found in maintenance manuals). No engineering change for this different type of motor (due to load requirements) had ever been submitted during the system design phase. The end result was incomplete documentation with no contract change being submitted to the client. Once the client followed the user requirements for lubrication, no further burn-out was ever noted.

Be prepared for those clients, purchasing agents, or individual contractors who never think about the important question "Who will fix it?" Discover maintenance requirements as early as a make-rather-than-buy situation. Determine the level of maintenance that will be needed. Obtain answers to the now-familiar questions:

Who will perform the maintenance?

What type of maintenance will be required?

Where will it be performed?

When will it be performed?

Why will it be necessary?

How will it be accomplished?

(Include the requirements of the manufacturer warranty, and also what maintenance the warranty states that the user **not** do initially.)

Use answers to the above questions to determine the class, or level, of maintenance to be performed (figure 4.11).

Plan for the level of maintenance to be recommended. Will the user perform it? Will the client plan for an in-house person to perform periodic maintenance and testing? Is an outside contractor's customer service representative needed? Will the client need to develop a maintenance team? Will the devices, units, or the total system ever need to be returned to the manufacturer or vendor? If so, will replacement items maintain customer conscientiousness? Note that these maintenance decisions in all end-of-phase reports are included in the documentation delivered to the client.

Ask how an item will be fixed during the Conception and Study Phases. In the Design Phase, the answers to the "who, what, when, where, why, and how" maintenance questions should become an integral part of the design process. Plan maintenance access to key parts of the system (including its software), and determine those points of access prior to manufacturing. This will save repair time and money at some future date. Be ready to present the client with procedures or, with large systems, total maintenance requirements, including specifications and maintenance manuals.

Use maintenance planning to impress a client, to enhance sales, and to capture repeat business. Team leaders must remember to budget for maintenance planning in addition to other normal expenses.

During the Design Phase, the solutions proposed and documented during the Study Phase are examined in more detail. Using the Design Description with Preliminary Sketches and the Design Specification as a guide, the team proceeds with developing and completing the design.

During the Design Phase, the solutions proposed and documented during the Study Phase are examined in more detail. Using the Design Description with Preliminary Sketches and the Design Specification as a guide, the team proceeds with the design. The tasks, schedules, and costs to construct the selected project are estimated and modified as a variety of design details are proposed and pursued. The chosen design details lead to a final design. Estimates are now as accurate as is possible.

Proposed project designs are circulated in the form of reports and drawings. More meetings are held to compare and evaluate the various proposed designs, and more meeting reports (section 8.2) are written. These reports and drawings eventually lead to three Design Phase documents. The documents completed during the Design Phase are the following:

1. The Letter of Transmittal or Contract Letter: the explanatory document
2. The Production Specification: the final specification
3. The Working and Detail Drawings: the drawings and plans

The content of each of these documents is described below. The flow from the Conception Phase documents and the Study Phase documents to the Design Phase documents is shown in figure 4.12.

The Letter of Transmittal or Contract Letter summarizes the obligations of the client and group implementing the project. It includes information on the nature of the project, costs, and payment schedules.

The **Letter of Transmittal** or **Contract Letter** contains both legal and technical terminology. Typically, its content includes the following:

1. The identity of the client
2. The identity of the group implementing the project
3. A list of the applicable documents (such as specifications)
4. The documents that describe the project and a scale model where appropriate
5. The means for ownership transfer
6. Warranties, actual or implied
7. The cost of the project to be implemented and a payment schedule
8. How task, schedule, and cost changes will be processed
9. The signatures of those persons representing the client and the team or its parent organization

Before it is signed, this document should be reviewed by both the client and the project manager, as well as by appropriate legal personnel. An example Contract Letter, one for the activity center project used in chapters 2 through 5, is given in appendix D.

The **Production Specification** content is described in detail in section 8.1 on page 205. A typical Production Specification contains the following paragraphs or groups of paragraphs:

1. Scope of the Specification
2. Applicable Documents
3. Requirements

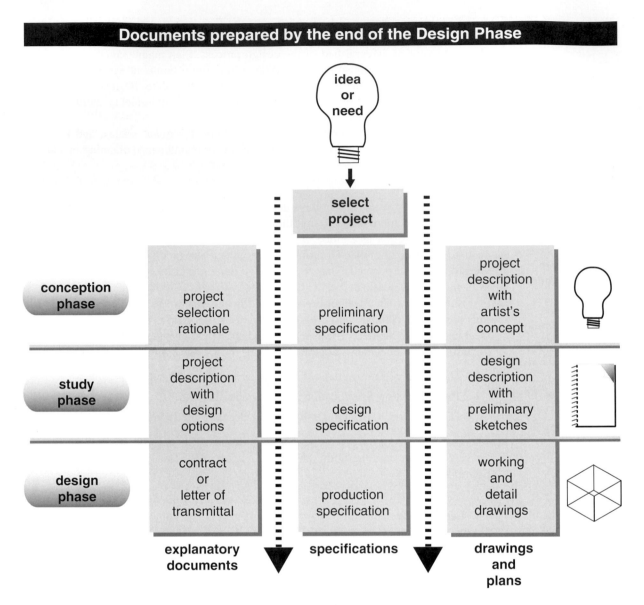

Figure 4.12   Documents prepared by the end of the Design Phase

4. Quality Assurance Provisions

5. Preparation for Delivery

6. Notes and Appendixes

Government specifications are more complex; industrial specifications are usually simpler.

The **Working and Detail Drawings** consist of the following:

1. A Presentation Drawing, similar to artist's sketches, that contains little or no structural information

2. Models where appropriate

3. Working Drawings, containing all necessary details

Depending on the product or software to be delivered, these drawings may require several pages or sheets or may require several volumes. When only experienced contractors are allowed to bid such a project, the detailed information may be significantly reduced.

SECTION 4.4   DEFINING PHASE COMPLETION

The Design Phase is considered "complete" when the documentation has been written, reviewed, and revised as recommended during the review. Again, the review team should include individuals from the funding sources and, wherever possible, experts in the field who are not responsible for the funding of the Implementation Phase. As noted in section 6.7, a designated person should continually monitor the expended costs and the projected remaining costs for the total effort.

The completed documents of the Design Phase consist of an expansion of documents originally devised during the Study Phase. These are the following:

- The Letter of Transmittal or Contract Letter
- The Production Specification
- The Working and Detail Drawings

Often, these three documents are initiated during the beginning of the Design Phase. They gradually evolve into their final content by the end of the Design Phase. See activity 3D on page 107.

Many project managers feel that the Design Phase is not complete until those individuals designated to be legally responsible for the work and its payment have signed the contracts.

Many project managers feel that the Design Phase is not complete until those individuals designated to be legally responsible for the work and its payment have signed the contract. Only then can the Design Phase be considered complete and the Implementation Phase initiated. Sometimes it is more realistic to terminate the Design Phase and reassign personnel until Implementation Phase funding is received. The project manager may save money in the short term. However, after a lapse of several weeks, obtaining the services of those who have previously worked on the project may not be possible. Then the cost of training new personnel represents a significant increase in the cost. So deciding when the Design Phase is ended may be a complicated decision.

Again note that, once the Implementation (Construction) Phase starts, any revisions to previous phases may be both expensive and time-consuming. Therefore, every effort should be made to have all designs and documentation completed during the Design Phase and before the start of the Implementation Phase.

CHAPTER OBJECTIVES SUMMARY

Now that you have finished this chapter, you should be able to:

1. Define the purpose and goal of the Design Phase of a project.
2. Explain the tasks involved in the Design Phase of a project and the activities or steps through which these tasks are performed.
3. Describe the role of a manager in the Design Phase.
4. Explain how a team should be organized for the Design Phase.
5. Outline a typical schedule for the Design Phase of a project.
6. Explain why human factors need to be considered in designing a product.
7. Describe how function and form are linked in developing design plans.
8. Define make-or-buy decisions, and explain how they are determined.
9. Describe the documents that should be prepared for the Design Phase.
10. List those who should participate in reviewing and approving the documents from the Design Phase.
11. Describe the documents prepared in the Design Phase and their uses.
12. Describe when the Design Phase is completed.

EXERCISES

4.1 For each of your options for a class project, revise and include the general Purpose and Goal statements given in section 4.1.

4.2 Examine the activities listed and described in section 4.2. Adapt them to each of your options for a class project.

4.3 Review the Study Phase effort with all team members; develop any required additional Design Phase details with the assistance of all team members. Organize the team into groups for the Design Phase. Assign each group a set of design tasks to perform.

4.4 Develop a design that will satisfy yourselves, your funding source, and the users of your product. Review the design with your previously identified judges. Evaluate their suggestions; decide which suggestions to incorporate into your design.

4.5 Study potential implementation solutions for the project design. Prepare the Contract Letter or Letter of Transmittal, Production Specification, and Working and Detail Drawings.

4.6 Contact potential vendors regarding make-or-buy information; determine make-or-buy decisions.

4.7 Review the Design Phase results and verify actual availability of Implementation Phase funding.

4.8 Refine the Implementation Phase tasks, schedule, and budget documents; plan the Implementation Phase. Include personnel assignments.

4.9 Verify that both your customer and your funding source will accept the Implementation Phase solution. Revise your solution as necessary.

4.10 Document the proposed project design and the reasons for its choice. Obtain Implementation Phase funding.

4.11 Describe how you would determine the optimum time to terminate the Design Phase. Terminate the Design Phase.

chapter **5**

1. What are the purpose and goal of the Implementation Phase of a project?

2. What tasks are involved in the Implementation Phase, and what are the activities or steps through which these tasks are performed?

3. How is money needed for projects typically obtained and delivered?

4. How are documents prepared during the Design Phase used in the Implementation Phase?

5. What tasks are typically part of the Implementation Phase, and how are they assigned?

6. What is the role of the project manager during the Implementation Phase?

7. Why are periodic meetings important during the Implementation Phase?

8. What are the common tasks that should be accomplished before actual implementation or manufacturing begins?

9. Why are prototypes developed?

10. What is the learning curve, and how does it relate to manufacturing production?

11. What documents should be prepared for the Implementation Phase and which should be delivered to the client?

12. What is "delivery" of a project, and what problems may occur at that stage?

13. What are supplier/manufacturer responsibilities and opportunities after the delivery of a product?

14. Why is having a review meeting at the end of a project important?

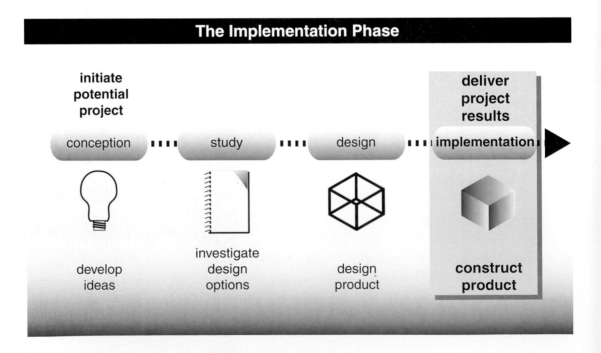

The reward of a thing well done is to have done it.
Ralph Waldo Emerson, 1844

The name for this phase—Implementation—implies that, after the three phases of conceiving, studying, and designing, it is time to construct and deliver your product or products. When this phase is complete, you will "have done it" (Graham, 1985).

SECTION 5.1

During the Implementation Phase, the design from the previous phases is converted into a finished product. Implementation costs are usually the greatest costs for the entire project.

IMPLEMENTATION PHASE PURPOSE AND GOAL

The **purpose** of the Implementation Phase is to provide the means, facilities, materials, and personnel for producing the end product. The **goal** of the Implementation Phase is to produce the completed design in accordance with the results of the previous phases. During this phase, you will continually review the product and its construction costs to ensure that they remain acceptable to the customer and the funding sources. Implementation costs are usually the greatest part of the costs for the entire project.

At the end of the Implementation Phase, you should know how to do the following:

- Review the Design Phase results after funding is received.
- Allocate tasks to team members and contractors as previously planned.
- Produce and control the quality of the previously designed product(s).
- Revise and prepare for delivery all documentation that will allow the client to operate and maintain the delivered product(s).
- Deliver the contracted product(s).
- Search for new work that will return a profit and enhance team skills.

Any changes offered during the Implementation Phase usually cause project delays and cost overruns. By now, no significant changes should be considered. You want to complete your project according to the specifications, on time, and within the money allotted.

As stated in chapters 1 and 2, the on-time completion of a project—according to specifications and within allotted funds—enhances the reputation of the manager, the team members, and the organization. This information quickly becomes known to those persons and groups who want quality work.

SECTION 5.2

IMPLEMENTATION PHASE ACTIVITIES

The activities for the Implementation Phase consist of the following:

Activity 1: Obtain Implementation Phase Funds

Activity 2: Review Design Phase Documents

Activity 3: Allocate Tasks to the Implementation Phase Team

    3A: Develop and monitor Implementation Phase details.

    3B: Organize the team into task groups.

    3C: Assign each group a set of tasks to perform.

Activity 4: Prepare to Produce the Designed Article

    4A: Define vendor schedules and deliveries.

    4B: Sign contracts with contractors and vendors.

IMPLEMENTATION

4C: Obtain applicable construction permits.

4D: Provide temporary on-site facilities, tools, and equipment.

Activity 5: Produce and Evaluate Product

5A: Fabricate and evaluate prototype model (if necessary).

5B: Fabricate the designed article.

5C: Produce and evaluate items to be produced in quantities.

Activity 6: Revise, Review, and Deliver Final Documentation

6A: Correct Working and Detail Drawings.

6B: Deliver Implementation Phase documents.

6C: Deliver warranties and associated invoices.

6D: Deliver final product and instruction manuals.

Activity 7: Operate and Maintain Customer-Contracted Items

Activity 8: Review the Results of the Implementation Phase

Activity 9: Pursue New Projects

9A: Prepare public relations material.

9B: Offer presentations to prospective customers.

9C: Construct speculative products.

Activity 9 is included for those organizations that wish to pursue follow-on business.

As was true for the previous three phases, the purpose and goal for the Implementation Phase lead to defining the activities for that phase. The sequence of activities given below may change from one project to another. However, all of these activities must be accomplished before the Implementation Phase can be considered complete.

After the contract is signed with the funding source and before construction is initiated, the Implementation Phase team must be assembled, tools and materials purchased, and the work space prepared.

After the prime contract is signed with the funding source (activity 1) and before construction begins, the Implementation Phase team must be assembled (activity 3B), work space prepared (activity 4D), and tools and materials purchased (activity 4D). The project flow is indicated in figure 5.1.

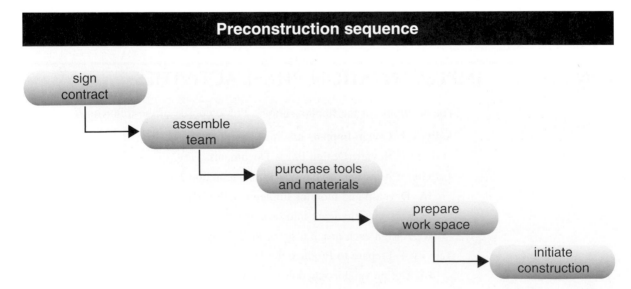

Figure 5.1 Preconstruction sequence

Activity 1: Obtain Implementation Phase Funds

Money is required to purchase parts and/or materials before construction starts. It can come from an advance payment provided by the customer or funding sources or from a loan given by a bank.

Money is required to purchase parts and/or materials before construction starts. If you do not receive an advance payment from the customer or funding sources, you may have to apply for a loan from a bank or other lending agency. Foundations usually donate all the necessary funds in advance of the Implementation Phase.

If the contract includes an advance payment, then you can deposit the money in a bank account to fund the cost of labor and the purchase of tools and materials. If the contract contains either a clause regarding **progress payments**—to be paid when a specified portion of the work is completed—or **payment upon completion,** then you may take the signed contract to a banker who will lend money against the contract. If a contract depends on a government or foundation grant, signing of the contract results in either a portion or all of the contract value being turned over to a bank or to the parent organization for the project. It then becomes the responsibility of a senior official in the government agency, foundation, or parent organization to monitor the progress of the contract and approve the disbursement of funds.

The signed contract may include or refer to the performance schedule and the specification. The project manager and each team leader should continually review these documents to ensure adherence to both costs and schedules.

Once the contract is signed, changes require extra time and effort. Team-initiated changes add to contract costs. These costs must be taken out of the contract profit, *if there is a profit.* A change initiated by the funding source must be paid for by the funding source. Such a change is known as a **change in** (contract) **scope.**

Unplanned changes that require additional money might occur during the Implementation Phase. There should be contingency funds to cover unplanned expenses.

Unplanned changes might occur during the Implementation Phase. There should be contingency funds to cover unplanned expenses. Banks, parent organizations, and foundations do not like being informed that more money is required to complete a project. Depending upon the wording of the contract, unused contingency funds may be either returned to the client, retained by the project, or included as a part of the project's profit.

Again, it becomes the responsibility of the project manager to analyze the effects of unplanned changes upon tasks, schedules, and budgets for an entire project and evaluate the trade-offs. The manager must select an appropriate course of action. The manager, in conjunction with a sales-oriented person, must convince the customer or funding source that contingency funds are important for flexible implementation. The customer or funding source should not have to be bothered each time an unplanned change occurs.

PROJECT EXAMPLE

The first meeting of the Implementation Phase team occurs on the afternoon of the first Monday in March. During the first part of this meeting, Professor Hulbert informs the team that the university has received the construction funds and has given him written authority to contract for the building foundation. The team can now order the materials based on the schedule previously devised. He reminds the team members that final exams occur during the third week in March. He also warns the team members that six months will pass rapidly. Roger schedules a meeting for Friday afternoon, at which he asks Alice, Joseph, and Manuel to contact vendors and obtain signed contracts within one week. (They may need to hand-carry paperwork in order to meet this schedule).

Activity 2: Review Design Phase Documents

Time may have elapsed while funding for the Implementation Phase was being approved. New team members need to be added and the selected contractors need to be informed and invited to participate. Each participant must now review the effects of any delay on the schedule.

IMPLEMENTATION

Both team members and contractors need to review the Design Phase documents so that they understand the tasks they are to accomplish, as well as cost and schedule constraints.

The schedules for delivery of work and materials may have to be revised. The selected contractors and subcontractors and specific members of the project team are assigned to perform the tasks that were carefully prepared during the Study Phase and revised during the Design Phase. However, because time has passed, some of the contractors and some members of the project team may no longer be available. Thus, assignments must be reviewed. Replacement individuals and companies must study the task descriptions and agree to their content. They must verify all of the following:

- What they are to accomplish (tasks)
- When they should start and finish each task (schedules)
- What their financial responsibilities are (budgets)

All team members and contractors should be given copies of the applicable task and schedule documents. Budgets of labor hours and materials costs should be accessible to each responsible team member. For small projects, *estimates* and *actuals* may be continually compared via wall charts. For larger projects, those persons in charge should be continually updated, perhaps by computer and Local Area Networks, to ensure that each group within the larger organization is staying within its budget and schedule as it performs the assigned tasks. As mentioned previously, the work may be coordinated by a separate group not directly involved in the construction (or fabrication).

Periodic project review meetings are often helpful. At these meetings, all those involved have the opportunity to meet and compare their ideas in detail on exactly how their individual tasks will be accomplished. The people involved must discuss and determine how various tasks are to be coordinated.

PROJECT EXAMPLE

It is now the first week of March. The ground in Bedford has not completely thawed. However, the foundation contractors assure Roger that this is not a problem: The foundation footing must be placed below the frost line.

The first meeting of the Implementation Phase team was on the afternoon of the first Monday in March. During the second part of that meeting, Professor Hulbert informs the team members that all orders for significant materials must be verified weekly to confirm that they will be delivered on schedule. He further reminded them that, with completion of the Design Phase the previous week, the activity center project is now scheduled to be completed in mid-September. The schedule cannot slip any further since classes for next year start the last week in September.

Activity 3: Allocate Tasks to the Implementation Phase Team

For the Implementation Phase, team members skilled in construction and manufacturing are usually added to the team.

The Implementation Phase normally adds team members who are skilled in construction and manufacturing. These performers are often referred to as blue-collar workers. These experienced individuals ensure that proper construction and manufacturing techniques are applied to the project. They are skilled in reading the Design Phase drawings and capable of implementing them.

Activity 3A: Develop and Monitor Implementation Phase Details

People with implementation experience enjoy building or manufacturing. They seldom need detailed guidance with regard to a particular task. However, since they may tend to focus on the details of their particular task, their efforts must be monitored to ensure that everyone's tasks are being implemented in the proper sequence and on a timely basis.

For all projects, implementation needs to be checked to verify that

1. Tasks are being accomplished properly

2. Schedules are being followed and achieved

3. Costs are being contained within their established upper limits

On small projects, the project manager is directly involved with day-to-day detailed activities and continually monitors the project's performance to verify that (1) tasks are being accomplished properly, (2) schedules are being followed and met, and (3) costs are being contained within their established upper limits. On larger projects, this function is performed by a person or group of persons in a project office. Monitoring and controlling a large project or program in great detail is a major effort.

Periodic review meetings are held to see if the project documents are being interpreted correctly. Changes may be considered during these meetings. Detailed documentation concerning any changes is prepared and distributed to all parties involved in the changes. Here again, it is the responsibility of the manager to approve, deny, or revise any proposed changes that will affect tasks, schedules, or budgets. No small detail can be neglected because it could affect the entire effort.

Before initiating activity 5 (Produce and Evaluate Product), the manager must initiate **management subtasks,** including the following:

- Assign project personnel to tasks

- Hire additional personnel as needed

- Hold project indoctrination meeting(s) for newly assigned personnel

- Sign contracts with selected contractors

- Notify materials vendors regarding start of contract

- Coordinate and obtain permits

- Acquire safety equipment required by OSHA:

    hard hats, safety ropes, safety glasses, and special tools

- Select and obtain work space furnishings, fixtures, and equipment

    provide first-aid kit

    obtain outdoor toilets

    rent or lease supervisor's trailer if necessary

- Purchase tools, construction supplies, and materials

- Occupy work space

Construction (or manufacturing) can now be started. Some planners prefer to identify the above set of management subtasks as a separate task. They include it as the first task on the Implementation Phase schedule prior to activity 5.

PROJECT EXAMPLE

At the Friday afternoon meeting, Roger describes the progress regarding site preparation. Roger states that he cannot assist directly in the initial construction effort, watch over implementation, and continue to achieve high academic grades. After a twenty-minute discussion, Alice is elected to become technical leader of the construction effort. Roger will continue to monitor the overall effort, including financing. Alice requests that each member note on the Implementation Phase schedule his or her assignment preferences. She gathers these notes and agrees to have initial assignments prepared by Saturday morning. Alice also agrees to visit the site and observe the start of the excavation. Ellery asks to go with her, and she accepts his offer.

Activity 3B: Organize the Team into Task Groups

The project manager must now accomplish the following:

- Hire semiskilled people to replace the creative people and the cost estimators.

- Discharge most of the creative people and some cost estimators, or pass them on to new projects. (See activity 9.)

- Conduct indoctrination meetings for all team members.

IMPLEMENTATION

When construction personnel are involved, they should be treated as part of the team, even though they may be contractors or subcontractors.

On Saturday, Alice examines the implementation tasks. She tentatively assigns team members to appropriate construction tasks and requests their comments. She then interviews potential semiskilled workers recommended by the contractors and suppliers who will be involved in the construction work.

Activity 3C: Assign Each Group a Set of Tasks to Perform

The project manager, with the assistance of others, converts all the documentation developed during the Design Phase into tasks for each team and contractor. (This conversion could be done during the Design Phase.) The project manager is responsible for making sure that each group understands how its assigned tasks interact with the tasks assigned to other groups and the contractors.

It is the second Monday afternoon in March. Alice explains the assignments and schedule shown in figure 5.2. She notes that Ellery, Kim, and Joseph will need some support during the framing and plumbing tasks. The remainder of the team offers to take turns assisting the assigned personnel, based upon their class schedules and obtaining transportation to Bedford. A bus schedule is obtained and distributed.

Chart of construction assignments: first two months

month	March	April
prepare site		
Roger		
Alice		
excavate site		
Alice		
Roger		
Ellery		
erect & cap foundation		
Alice		
Roger		
Ellery		
erect rough framing & plumbing		
Ellery		
Alice		
Kim		
Joseph		

Figure 5.2   Bar chart of construction assignments: first two months

Activity 4: Prepare to Produce the Designed Article

The Implementation Phase has been initiated and funds are available that can now be committed. Vendors must be informed, and their delivery dates must be confirmed. Construction permits, requested in advance to allow the town committees to review and comment on them, must be obtained before the start of construction. On-site temporary facilities for the construction workers must be provided. Now construction and/or production can begin.

Activity 4A: Define Vendor Schedules and Deliveries

Types of vendors include these:

- Profit-making companies that are incorporated
- Groups and individuals "doing business as" (DBA) themselves
- Independent contractors who prefer not to incorporate but to work as sole proprietors
- Nonprofit organizations, such as a vocational school

Various state and local agencies and departments watch over each type of vendor to ensure that vendors perform only that type of work for which they are licensed.

Once implementation starts, vendors must be contacted to determine if they are still available at the price they bid.

Contractors providing labor, materials, and expertise must be contacted (by telephone, letter, fax, or e-mail) to determine if their bid is still in effect—and at their quoted price. Vendors providing supplies and services must be contacted to determine if they are still available at the price they bid. It is also vital that labor, materials, supplies, and services be provided on time and at the quoted prices so the project can be completed on schedule and within budget.

PROJECT EXAMPLE

At the end of the second Monday meeting, Alice asks Manuel to assist Joseph in verifying shipment of the framing lumber and plumbing supplies. Manuel and Joseph are also assigned to contact Jim Sullivan to verify timely assistance from his plumbing volunteers and Jim Stander to confirm assistance from the electrician volunteers. All team members leave a copy of their expected spring (April, May, and June) class schedules with Alice for consideration in later assignments.

Activity 4B: Sign Contracts with Contractors and Vendors

Contracts identify, in a legal manner, the participants in the project, and the respective responsibilities of those participants. The contractor and vendor contract or purchase order typically contains the following information:

- Responsible persons: their names, organizations with which they are affiliated, and addresses
- Contract costs and authorized signatures
- Applicable reference documents and warranties
- Material or services to be provided
- Schedule of performance and progress payments where applicable

The signatories to contracts include the funding source (optional), the prime contractor, and subcontractors. (See appendix D.) Vendors usually receive purchase orders from the prime contractor or a subcontractor.

IMPLEMENTATION

It is the second Friday afternoon of the Implementation Phase. The North-eastern University treasurer meets with Professor Hulbert and those contractors and vendors who can attend. The participants sign, and are given copies of, their contracts. The lumber vendor brings sandwiches for all and invites them to tour the lumber yard. The vendors who could not attend this meeting mail their signed contracts. These arrive in the Monday mail.

Activity 4C: Obtain Applicable Construction Permits

For projects involving work on land and buildings, construction and other permits must be obtained in a timely manner so that they are available when construction is ready to start.

For projects involving work on land and buildings, construction and other permits must be obtained. The length of time required for processing applications for these permits must be known in advance because any delays will cause work slippage. Thus, the project manager or other team member assigned to the task must request applications from the appropriate licensing boards or other governmental agencies in a timely manner. In this way, the actual permits will be available when needed.

In some situations, the government officials involved must inspect the property prior to the issuance of permits. They must also inspect the work in progress and perform code inspections to certify that the work is being performed according to the applicable government rules and regulations and the information stated in the permits.

Many of these rules and regulations overlap or require interpretation. Thus, it is necessary for the project manager to coordinate the actions among the various government agencies because overlapping authority and disagreement may exist. At times, face-to-face meetings must occur. Then the various government representatives can agree on an acceptable solution to conflicting requirements, and the project can continue.

Fortunately, during the early days of the previous December, all of the permit forms required had been obtained by Alice and Jose. They were immediately prepared, verified in content, and submitted to the appropriate government agencies and licensing boards. The necessary application fees were paid at that time. Now the approved permits are issued.

The Northeastern University treasurer had the checks prepared in advance for the permits so work could progress on the planned schedule. (The electrical and plumbing permits will be obtained later by Jim Stander and Jim Sullivan, who will be supervising and assisting in that work.)

Activity 4D: Provide Temporary On-Site Facilities, Tools, and Equipment

Temporary facilities, such as work space, office space, and storage space, must be provided at a new construction or manufacturing site.

Temporary facilities, such as work space, office space, and storage space, must be provided at a new construction or manufacturing site. Portable toilets and first-aid facilities must also be provided for the workers. Furnishings and fixtures for use by contractors, such as desks and chairs, are necessary, and they should be included as a part of the cost and schedule estimates developed during the Design Phase. OSHA-required safety equipment must be determined and purchased.

Also, tools and other supplies required by the project team must be ordered and purchased; these should be so noted in the task descriptions and accompanying budgets. Temporary telephones and electrical power connections must be provided via the local utility companies.

Decisions regarding renting, leasing, and purchasing the above items are a part of the Design Phase planning for the Implementation Phase. Any last-minute changes in these plans must be made before construction and manufacturing can begin.

Note: If management tasks and subtasks are extensive, then they should be inserted here. See page 127.

> Alice's uncle loans a small trailer-office to be used on-site for supervisory personnel. An electric power line is connected via a meter on a pole. The power line is connected to a circuit-breaker box with a disconnect switch and a pair of 115-volt, 60-Hertz outlets. The town provides a temporary water connection. A telephone is installed, and a portable toilet unit arrives.
>
> Roger verifies, via Alice, that all concerned have a copy of the tasks, schedules, budgets, and the Specification for the activity center (see appendix C). Ellery is assigned the responsibility for distributing drawings to the appropriate team members and contractors. The university loans a computer so that Jose and Kim can update the schedules and budgets on site.

Activity 5: Produce and Evaluate Product

For complicated devices and software, proof of performance may be accomplished by first constructing one item, known as a prototype.

Finally, the time has come to actually start construction. All of the planning must now result in a quality product, constructed in a timely manner, in accordance with the Design Specification, and within budget. Every project will be different; some of the problems encountered are included in the activity center example (see activity 5A below).

Manufacturing occurs when quantities of the same product are produced (Kaplan, 1993). For complicated devices and software, there are times when proof of performance must be accomplished by first constructing one item (activity 5B). That item is often referred to as a prototype of the production units that will be fabricated later, once tests of the prototype are successful.

After prototype construction, evaluation, and revision where necessary, production can begin. The first production models (activity 5C) are evaluated to ensure that all new production tooling and procedures are performing as designed. Larger production items are tested as subassemblies prior to final assembly.

Activity 5A: Fabricate and Evaluate Prototype Model (if necessary)

Prototypes may be constructed (1) to verify design assumptions or (2) to test production techniques.

Activities 5A and 5C apply to the fabrication and evaluation of several items of the same design. These activities do not apply to the activity center, unless the team decides to construct a scale model.

For production items, prototype construction may be required for either one or both of two major reasons:

1. Design assumptions and costs may need to be verified.
2. Production techniques may need to be tested.

Sometimes these two "proof via prototype" reasons occur sequentially (figure 5.3). When the costs of production tooling are excessive, then the production techniques are improved via the first production models, as discussed next and in chapter 9.

Prototype construction provides a model that allows for verification of the design drawings and accompanying specifications. Human factors—how humans and the product (hardware or software) interact—can also be tested. How the operators behave under stress or when they are tired can be determined by simulating problem situations and applying them in a sequence that causes operators to experience stress and fatigue.

Prototype units can also be tested in laboratories under simulated operating conditions to verify that they meet performance specifications. Environmental tests include shock and vibration, also known as "shake, rattle, and roll" testing. Temperature and humidity variations can also be simulated; salt-spray testing—to determine resistance to corrosion—may be performed.

Field conditions may be simulated by using the prototype unit in a location that closely approximates the environmental extremes to which the production units will be exposed. It is not possible to test for every variation that can occur. However, experienced professionals can often predict what may occur during extreme conditions by extrapolating from the test results.

IMPLEMENTATION

Figure 5.3 Prototype development and evaluation

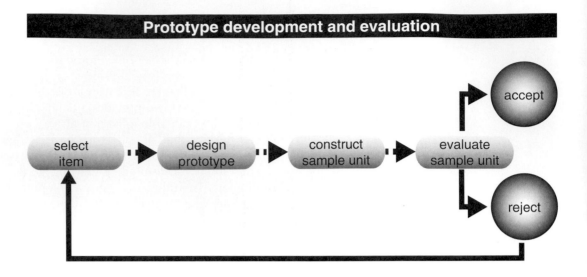

Figure 5.3 Prototype development and evaluation

Whenever possible, a production prototype is constructed using the same or similar techniques, equipment, and processes that will be used in final production.

Whenever possible, a production prototype is constructed using the same or similar techniques, equipment, and processes that will be used in final production. This approach may be expensive and/or time-consuming. Thus, decisions regarding the techniques, equipment, or processes to be used in final production require the advice of experienced manufacturing personnel.

Activity 5B: Fabricate the Designed Article

As the product is about to be fabricated, the physical conditions needed for efficient production must be reviewed.

Fabrication of a product may require preparing a site or building space. Room must be provided for the work about to be accomplished. Revisions to the existing site must occur before fabrication can begin. Environmental impact may need to be assessed, and measures to protect the environment may be required. Material storage areas may be needed. Security needs will have to be assessed to eliminate—or at least control—theft.

The workers involved on a project vary greatly in their need for supervision. Some workers have performed the same type of work so many times that they may need only general supervision. However, with the advent of monitoring environmental impact upon the location, additional training of these workers on environmental issues may be required.

As work progresses, the designated tasks are performed. Budgets and schedules are closely monitored to ensure that the Implementation Phase plans are either followed carefully or modified to fit the actual need.

PROJECT EXAMPLE

The site for the activity center has been prepared by town of Bedford employees, who have removed the topsoil and stored it locally. With the assistance of a third-year civil engineering student, Joseph and Manuel survey the foundation area, define the location of the foundation corners, and install the batter boards. (See appendix B.) Then the foundation contractor arrives with the excavation equipment and digs the hole for the foundation. Fortunately, no ledge is encountered.

The forms for the foundation spread footings are then constructed and braced; the concrete slump is tested; the footing is then poured. A few days later, the foundation and chimney forms are installed. The foundation concrete is slump-tested and then poured. Plastic drain pipe and pea stone arrive and are placed around the foundation footings. The pipes are then connected to an existing storm drain.

A few days later, the foundation forms are removed. Six inches of hay is firmly packed on top of the pea stone by Kim. After the foundation outer walls appear to be dry, two coats of asphalt emulsion are brushed onto the foundation by Laura; she creates a design that, sadly, disappears when the soil is returned to "backfill" around the foundation. It is compacted by Roger and Jose after school. The rough grading around the foundation is complete. Kim is also learning how to update the schedules and budget spreadsheets.

Alice observes that some of the town laborers assisting at the site exhibit crude manners, saying things and using language that offends the students. She discusses this problem with her Uncle Bill at Sunday dinner. He asks her, "Does the problem seem to be gender- or race-related?" Alice replies that it seems to be more a problem of blue-collar workers versus students. He comments that his workload in nearby Stoneham will not totally consume his days. He volunteers to visit the site most afternoons to review progress and handle problems. Alice and Bill call Roger and Professor Hulbert, who accept his offer.

Uncle Bill usually arrives in the late afternoon to inspect progress and answer questions regarding the next day's work. For example, the lally columns are late in arriving. A quick telephone call from Uncle Bill ensures their prompt delivery.

It is now early April. Most of the lumber arrives and is stored nearby. (The local police watch over it during nights and weekends.) The foundation cap is installed. The main carrying timbers are installed on the lally columns. The main floor joists are prepared and installed; the plywood subfloor is then installed by the team in one weekend.

The concrete trucks return; the floor slab is poured and leveled, and a floor drain is installed. Uncle Bill shows the students how to verify the level of the floor as it is poured. He privately talks to some of the laborers, explaining that the students are here to both work and learn. He encourages them to be understanding of these young students who are just now acquiring some practical experience. He also talks to the students about respecting the laborers, who are skilled in their occupations. He reminds both groups that they are all human beings who deserve to be treated with respect.

It is now mid-April and the construction is approximately one week behind schedule—partly because of weather and partly because of mid-term exams. Roger and Alice decide to ask members to work at least the next two weekends—weather permitting. The days are getting longer, daylight saving time has begun, and the delivered lumber is relatively dry. On the first weekend, Uncle Bill arrives with some special Italian food that he has prepared. He offers advice on how to install corner bracing. That day all of the team members learn much and eat well. The house is framed, which includes the construction of main floor timbers, rough flooring, wall studs, door and window openings, exterior wall plywood, ceiling joists, and roof rafters.

Laura is a little nervous regarding schedule slippage and cost control. On Monday, the team members meet and agree on how to increase their work output and how to save some money as well. Roger agrees to call those vendors who are late in delivering materials. They improve their deliveries so that work can gradually return to the original schedules.

Plumbing supplies are purchased from a distributor, thus saving a few hundred dollars when compared with the originally estimated retail prices. Rough plumbing is installed on the last weekend of April, when a licensed plumber (Jim Sullivan) can be there to both guide and assist the workers.

It is now the second week in May. The roof is framed and boarded with sheets of CDX exterior plywood. The roof and chimney are one week late being constructed—again because of weather. The brick mason agrees to work one weekend at no extra cost. She installs the chimney and roof flashing as part of constructing the chimney. Jose assists her by carrying the bricks to where she is working; Manuel assists by mixing the mortar under her guidance.

IMPLEMENTATION

Kim is now looking for ways to get the project back on schedule. She chairs the next Monday meeting and offers her suggestions. The doors and windows are expected to arrive two weeks late. Alice and Joseph agree to contact the vendors to see if the delivery dates can be improved. They discover that, because of the slippage of another project, the windows and doors can now be delivered on schedule.

Uncle Bill arrives just in time to correct the team members' approach for installing the prehung windows and doors. He saves the team members a few days by assisting on the following weekend and, of course, by bringing some more excellent food. During lunch, he explains to them how work can be accomplished with less effort and performed more expertly. Laura decides to write a term paper that describes his suggestions. The exterior finish is installed; that effort is followed by the installation of the exterior siding, which consists of cedar shingles.

Jose now works with a local electrician (Greg Roberts, who is assisting Jim Stander). With Laura's help, they install the rough wiring in one week, including one weekend. The shingles are also coated with bleaching oil that same weekend. Now the project is back on schedule. Manuel is delighted and issues the corrected schedule.

The rough plumbing is installed with the help of one of Jim Sullivan's sons. The roof has to be penetrated for the bathroom and kitchen vent stacks. This work is accomplished on schedule.

It is the middle of June. The insulation and Sheetrock™ have arrived, and it is exam week. Thus, the materials are unloaded and stored within the center, since it can now be locked at night. Two additional staple guns and face masks are borrowed. Following exams, all team members work on installing insulation and Sheetrock™ so the schedule will not slip. When approximately one half of the insulation and Sheetrock™ is installed, the team is split into two groups. Jose plans to paint, so he is placed in charge of taping and spackling (covering the joints with compound) to provide smooth wall and ceiling surfaces. He then works and also supervises the sanding and interior painting. To stay on schedule, all team members work on the Fourth of July. They are invited to stay and enjoy the Bedford fireworks that evening. Much watermelon is consumed.

Roger and Alice have noted some friction among group members. Professor Hulbert reminds Roger and Alice that no one has had the opportunity to enjoy a week away from the effort; staggered vacations are then discussed and planned so everyone can leave for a week and return refreshed.

It is now mid-July. The flooring has arrived one week early; it is stored in the basement. The interior painting is begun, followed by installation of the flooring, room by room, after the paint has dried. This effort progresses slowly and is not complete until late August.

Roger discovers that the total charges are beginning to exceed the funding. He has to acquire an additional $2 000 quickly. The team decides to have a Labor Day open house for the townspeople. They leave a basket at the door and collect $1 850. Professor Hulbert, Bill D'Annolfo, and Jim Sullivan each contribute an additional $50 to the fund.

The first of September and Labor Day pass quickly. The electrical fixtures arrive on time and are installed on schedule. The plumbing fixtures are late in arriving, and their installation occurs after Labor Day, yet the installation is completed on schedule.

The New England weather has tuned cold early this year. Uncle Bill suggests that, to avoid further schedule slippage, the final landscaping should be initiated now. So, grass seed is sown and watered. Accomplishing this work becomes a problem because of preplanned vacation schedules.

It is now the middle of September. The beginning of the fall quarter of classes is only one week away. Professor Hulbert and Uncle Bill inspect the activity center to be certain that nothing has been forgotten. A few interior corners are discovered to have been painted poorly. Three of the siding shingles were not bleached. Some of the grass is not growing; those areas need to be reseeded. Alice calls on the few team members who have returned to school early to assist in performing the necessary work so a formal delivery date can be firmly established with the town officials. The corrections are completed during the next ten days.

Activity 5C: Produce and Evaluate Items to Be Produced in Quantities

The learning curve refers to the ability of operators to gain in productivity as they learn how to use tools more efficiently and as they gain dexterity.

The production of many identical items requires additional planning. The first few models require more time to manufacture; later ones benefit from the learning curve (figure 5.4). Production specialists are responsible for including learning-curve information in their Design Phase estimates of production costs.

A learning curve can be graphed as a curve of productivity versus time. When an individual is starting to learn to operate a machine or to perform an assembly operation, the level of performance is low at first until level A is reached. As the operator gains experience, the level of performance increases from level A to level B. This change is caused by increased experience and familiarity with the tooling and processes. Further experience gradually improves the rate of performance from level B to level C. This change is caused by increased dexterity. Additional experience will lead to a minimal improvement in performance from level C to level D. After level D, the improvement in performance is negligible. If greater productivity is desired, a change in manufacturing technique, procedure, or equipment is required. When prototypes are built, different manufacturing techniques can be tested and evaluated to determine the most productive ones.

The last tasks of a manufacturing plan involve inspection, testing, and shipping.

The last steps in manufacturing items involve inspecting, testing, and shipping them promptly.

1.   *Inspection* is usually done near the production facilities so that defective items can be quickly returned to the production group for correction and adjustment. For simple items, only selected samples are tested; however, for complex items, each manufactured example is thoroughly tested. The tolerances, which are the deviations allowed from the design-specified values, are established during the Design Phase. Measuring equipment must be available to check any variations from the tolerance values specified. The selection of inspection and test equipment is done by quality control specialists.

Figure 5.4   The learning curve

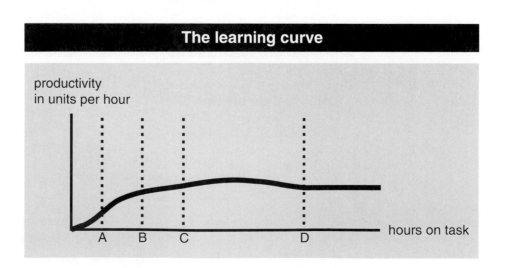

2.   *Testing* may involve using equipment to verify that it meets all performance specifications. Pilot units may be used to establish that the manufacturing personnel correctly fabricate and assemble the equipment.

3.   *Shipping* is an additional factor to be considered in preparing products. For small, delicate items, shipping materials are available to protect them from being damaged in transit. For large, delicate items, crates are often used to protect the equipment. Recording devices can be installed in these crates to record the time and amount of shock and vibration so that the location where damage occurred can be established (since time can be converted to location).

The Implementation Phase may apply to units, devices, or systems and procedures. Where large-scale manufacturing is involved, the Implementation Phase is often divided into three subphases shown in figure 5.5:

1.   *Construction: Fabricate sample items and evaluate results.* Only one or a few items may be constructed. These items are normally referred to as prototype models or pilot runs.

2.   *Production: Manufacture a significant quantity of items.* Several or many items may be constructed. A manufacturing line may be designed and established to produce large quantities of the chosen item in order to apply the learning curve concept to cost and schedule efficiencies.

3.   *Employment: Operate and maintain items.* A simple example of equipment employment is a copier machine. The equipment manufacturer and/or copier seller perform both preventive and emergency maintenance. These machines may be leased or purchased from the manufacturer's or distributor's sales office.

Some engineering texts consider employment to be a separate phase following delivery. This makes sense because different parts of the company, or different companies, may provide the operation and maintenance services. (See activity 7.)

PLANNING AHEAD

Production tooling is evaluated through exhaustive testing of a significant quantity of the first items off the production line. A balance must be achieved between the cost of this testing and the loss of reputation and business caused by products that do not meet specifications or customer expectations. The manufacturer may have placed higher expectations in the mind of the customer, via product marketing and advertising, than the product is able to deliver. Thus, there must be significant coordination between marketing and engineering if the customer is to receive what is advertised.

Automobile recalls are a result of insufficient product testing under very specific conditions. The incorporation of the technique known as "total quality control" greatly reduces manufacturing errors. The technique stresses the importance of doing things right the first time. Total quality control must be included at the beginning of the design process to reduce the occurrence of design and production deficiencies.

Figure 5.5 Implementation subphases for manufacturing

Implementation subphases for manufacturing

construction subphase
fabricate one or more items

production subphase
manufacture a quantity of items

employment subphase
utilize a product in a useful way

It is desirable to minimize the need for storing and maintaining large inventories. Large inventories require increased space and additional financial investment. A scheduling technique known as "Just-In-Time Performance" can be applied. This technique provides for shipping the needed components at a rate just equal to the rate of consumption. Only a very small inventory is required to continue the operation of the assembly line in a smooth and effective manner.

Activity 6: Revise, Review, and Deliver Final Documentation

During implementation, documentation corrections are often neglected. Sometimes changes are decided at the construction site and are approved verbally. People are busy and may neglect to prepare a formal change notice. However, any documentation to be delivered to the customer must represent the actual, final product. Therefore, all documentation should reflect the changes made at the site.

A portion of the delivery may contain material that has been purchased from a vendor and that is under warranty. That warranty must be passed on to the customer so that the customer can directly contact the vendor without bothering the project's parent organization.

Activity 6A: Correct Working and Detail Drawings

A Change Notice is prepared when changes are being made to the original design plans during implementation; these changes must be documented.

A Change Notice is prepared when changes are being made to the original design plans during implementation; these changes must be documented. This notice is sent to all those involved in design, construction, or manufacturing, and to the person or group in charge of documentation. It may be initiated by the design group, the construction organization, or the manufacturing organization. A meeting of all involved personnel may be necessary to explain the need for the change and to further explain how the change is to be implemented.

*All those involved should participate in the discussions about changes so each person (or group of persons) realizes the effect of changes upon the project. This is known as **bottom-up** management.*

During construction, changes may be required that could not be predicted in advance. Budgets may have to be changed; however, the total dollars should be kept close to or within the contracted amount. Tasks and schedules may have to be revised. These changes are the responsibility of the project manager and, in much larger projects and programs, the project control people. Whenever possible, all those involved should participate in discussions. In this way, all persons (or groups) realize the effect of changes upon their part of the project. This is known as **bottom-up** management, which is discussed in section 6.3.

PROJECT EXAMPLE

Ellery is assigned the task of documenting any changes to the various drawings as they occur. He quickly realizes that the changes need only be inserted into the drawings via the CAD program once or twice each week. He also discovers that sometimes a change may cause a previous change to be canceled. He informs all concerned individuals that a set of the most recent drawings is on display on a table in the loaned trailer. All proposed or processed change orders are available on a clipboard next to the set of up-to-date drawings.

Appendix E contains an example of a Contract Change Notice (CCN) developed specifically for the activity center project. The fifth change was requested after Robert Lanciani, an acquaintance of one of the students, happened to visit the construction site. As a trained and experienced firefighter, he noticed that the metal Simpson TP gusset was indicated on the drawings and in the Specification for the Activity Center (appendix C). At a recent fire where the heat (but not yet the fire) had reached the roof, the heated metal gussets fell out, causing the trusses to fail. From this experience, Lanciani pointed out that metal gussets and the associated trusses were a danger to firefighters. Therefore, at the activity center construction site, an informal conference occurred that led to a change order (noted in appendix E).

IMPLEMENTATION

Activity 6B: Deliver Implementation Phase Documents

For the Implementation Phase, the completed documents to be delivered to the clients for their use are (1) instruction manuals, (2) warranties, (3) maintenance specifications, and (4) As-Built Plans. The content of these documents is described in section 5.3.

A package of the corrected Working and Detail Drawings (activity 6A) has been gathered together. It must be accompanied with written material that (1) describes what changes occurred during construction (or manufacturing), and (2) explains why these changes occurred. Thus, both the client and the funding source(s) will have documentation that is correct and up-to-date. Included are explanations regarding the changes that occurred during the Implementation Phase.

Examine all change notices and incorporate information on the revisions in the documentation being prepared for delivery to the client and/or funding source(s). Review all construction or manufacturing drawings for changes that were not properly documented with change notices. Insert these changes into the proper place in the overall documentation. It may be necessary, during more formal and large projects or programs, to issue additional change notices upon construction completion to document all changes.

PROJECT EXAMPLE

Ellery has been given the assignment of noting changes to the specifications. As with the Working and Detail Drawings, he informs all concerned individuals that a corrected copy of the Specification for the Activity Center is on display on a table in the loaned trailer, next to the drawings.

PLANNING AHEAD ▶

Activity 6C: Deliver Warranties and Associated Invoices

Promptly after receipt of goods, you must obtain warranties and copies of the applicable invoices that verify proof of purchase by you or your team. Thus, if an item that you have purchased fails, it can be repaired or replaced by the original source or manufacturer—usually without an extra cost to your team.

For companies that look forward to "repeat business," as well as for good recommendations from their current customers to attract new customers, it is desirable to ensure customer satisfaction. This can be accomplished by trying to meet the following criteria:

- Support warranties faithfully
- Perform design revisions that improve the product
- Prepare service contracts for product maintenance

A satisfied customer usually speaks well of the team who performed the work; that customer will recommend the team to others. If modifications to the equipment or software lead to more use of product and more flexibility in its use, follow-on contracts are likely to occur.

Many manufacturers provide servicing of products. This work may be covered by a service contract. Such a contract may cover both periodic and emergency maintenance and repair.

Warranties should be clear and specific. The equipment now belongs to a user who may not want to be concerned with its maintenance and repair. For continuous and reliable equipment performance, most users are willing to purchase a service contract from the manufacturer rather than having to maintain that equipment within their organization.

PROJECT EXAMPLE

As construction progresses, Kim is assigned the responsibility of gathering together all warranties and instruction manuals and filing them for later delivery to the town of Bedford. She soon learns to also obtain and copy the applicable invoices before they are sent to the university for payment. The invoice copies then become a part of the documentation package to be delivered to the town of Bedford.

Activity 6D: Deliver Final Product and Instruction Manuals

Delivery is considered to be an instant of time at the end of the Implementation Phase. There are three conditions associated with delivery:

1. Delivery may be refused by the customer until all questions and misunderstandings are resolved.

2. Delivery may be "conditional" while customer-requested revisions are being completed.

3. Delivery may occur with acceptance by the customer.

Delivery formally occurs when the applicable documents are signed. The associated paperwork transfers the responsibility for, and ownership of, the project to the customer for whom it was produced.

Delivery formally occurs when the applicable documents are signed. The associated paperwork transfers the responsibility for, and ownership of, the project to the customer for whom it was produced. This delivery and acceptance should include payment, or an agreement to pay, for the work satisfactorily accomplished.

Where manufactured products are involved, the customer may inspect the product at the customer's plant. This function is known as incoming inspection.

When the product is either large or complicated, then the customer may prefer to have a quality control inspector at the manufacturer's plant, as shown in figure 5.6. This is known as "inspection at the vendor's facility." Thus, necessary changes and/or repairs can be performed at the vendor's manufacturing facility without involving extra shipping costs and significant time delays.

It is very important to realize that the customer often accepts individual items or portions of a project as they are completed. Customer reviews may also have been a part of the design and development process. However, when it is time to deliver the product(s), the results may not be acceptable to the customer. Unplanned changes may be required to satisfy the customer.

A structure, equipment, or item of software often appears different once it is completed, even if the item accurately implements the drawings and specifications. Some people have a difficult time visualizing an outcome. They may have an idealistic picture in their mind regarding results—fantasy versus reality. For these customers, scale models and prototypes can be helpful.

PLANNING AHEAD

Those persons accepting delivery may believe they were going to receive something different from the item(s) constructed. The project manager must apply personal skills to convince the customers that they have received the items for which they contracted. The alternative is a legal battle that may consume considerable time and funds. This is another reason why it is important to continually involve the customer in the development of a project.

Figure 5.6   Delivery and acceptance

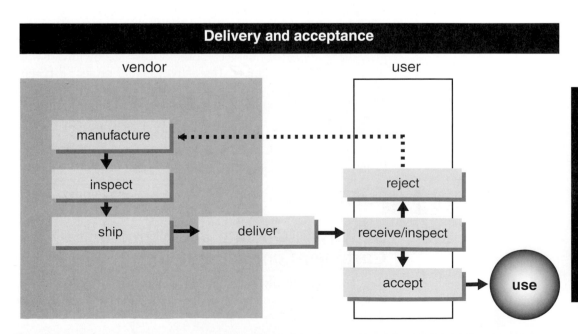

IMPLEMENTATION

Prior to asking a customer to accept a project, one or more team members must review the plans and specifications to ensure that the completed project actually complies with them. They must place themselves in the shoes of the customer. In this manner, the team is trying to avoid surprises when it requests customer acceptance. Persons who represent the customer may be involved because they will later be accepting responsibility for the finished product or project.

Delivery may have occurred satisfactorily. However, on initially working with the delivered project or items, the customer may decide that some portions are not performing as expected. Whether the contractor should assist in the desired (or even demanded) changes at no cost is a difficult decision. Some loss of profit may be less costly than an extended legal battle. There is also a possible loss of follow-on business as the result of a poor, even if undeserved, reputation for the project team.

Examine the following unfortunate example: The prime contractor for a military mobile digital computer system designed and tested the computer. Another company, under a separate government contract, designed and constructed the tape drive to be used with that computer and with other government computers. Contractually, the federal government accepted responsibility for the two separate specifications. However, when the computer and tape drive did not operate together properly, the federal government blamed the prime contractor. The prime contractor then had a very difficult time obtaining contracts to supply mobile computers to other federal agencies. The prime contractor was perceived to be at fault because of the failure of the computer and the tape drive to operate as one system. Thus, perception and reality may be in conflict. Someone must decide: Is the customer always right? and How will other potential customers perceive the problem?

Customer satisfaction can lead to more opportunities for an organization that wants to be profitable and grow in service and reputation.

Customer satisfaction can lead to more opportunities for an organization that wants to be profitable and grow in service and reputation. Recall that any project or program, from selection to delivery, consists of four relatively distinct phases: Conception, Study, Design, and Implementation. Note that, in each phase, documentation has been stressed. This helps to ensure that there are no unhappy surprises.

Lawyers are retained when people do not agree on the interpretation of documents. Therefore, it is vital to prepare brief, simple, yet complete documentation so that a project can be successful, satisfying all those involved. The funding sources, as well as those who will formally receive delivery of the project and its documentation, must especially be satisfied with the results.

PROJECT EXAMPLE

Mr. Williams, who is superintendent of the Department of Public Works (DPW) for the town of Bedford, visits the site of the activity center from time to time. Roger and Alice soon learn that at least one of them should be available to answer any of his questions. At each visit, he is shown the changes to the drawings and specifications so that the team is assured that they will be acceptable to him at "delivery time." He advises team members on how he would prefer delivery of the warranties and accompanying invoice copies for easy use by town employees.

It is late September. The activity center project is complete. Building inspectors perform the final inspections, verifying that the completed structure conforms with all applicable local, state, and national building codes. The team receives a verbal commendation from the inspectors for a job well done.

A plaque is designed with the names of the Northeastern University team members, the activity center benefactors, and the town officials. A friend of Ellery's offers to prepare the plaque in time for the formal activity center delivery. Mr. Williams is called and the date for formal delivery of the activity center is established.

Activity 7: Operate and Maintain Customer-Contracted Items

The responsibility for maintenance and repair of one item or product (such as an aircraft) or many items (such as copier machines) becomes an important income opportunity. If a group of students from a technical school, college, or university construct an item, then it must be agreed in advance that the educational organization will not be held responsible for maintenance and repair. (Why? Once the individual students have graduated, their skills and knowledge are no longer available.) For situations in which an existing company wishes to service its products, then a separate maintenance division or a separate service company may already exist or must be formed.

Maintenance is grouped in two categories:

1. Routine preventive maintenance, to ensure that the equipment operates as designed and that worn parts are replaced before they fail
2. Random maintenance, which cannot be predicted in advance but can be estimated on a statistical basis

A contractor can plan for both preventive maintenance and random maintenance, and it can estimate the number of employees and supervisors required.

A contractor can plan for both preventive maintenance and random maintenance, and it can estimate the number of employees and supervisors required. However, there will be times when two or more random failures will occur at the same time. A good reputation is earned when effective routine preventive maintenance prevents such simultaneous multiple random failures. This is *not* an easy assignment for a firm!

Activity 8: Review the Results of the Implementation Phase

*On completion of the work, all involved individuals should meet and review the work accomplished and the problems that arose during its implementation—known as **lessons learned.***

When people are busy working on a project, they may miss some important learning opportunities. "It is difficult to see the forest for the trees." On completion of the work, all involved individuals should meet and review the work accomplished and the problems that arose during its implementation—known as **lessons learned.** Such a meeting should not be an opportunity for individuals to blame others for the problems that had to be solved. Rather, the group should focus on what to do differently so the next project with which they are involved will flow more smoothly.

PROJECT EXAMPLE

On completion of the project and prior to formal delivery to the officials from the town of Bedford, Professor Hulbert conducts a meeting with all the team members present. He has each one describe what he or she has learned from working on the project. The list on the chalkboard grows, with some laughter as team members realize how different reality can be from theory. Northeastern's President Curry then arrives and presents each team member with a certificate of accomplishment from the university.

Sandwiches and soft drinks are served; the meeting is formally adjourned by Professor Hulbert as he reminds students to concentrate on their studies so they can graduate successfully four years from now. He reminds them to prepare resumes that include their work on the activity center and to hand-carry a copy to the Cooperative Education Department, which helps students obtain "co-op" jobs as part of the school's work/study curriculum.

PLANNING AHEAD ▶

Activity 9: Pursue New Projects

The pursuit of new projects applies to organizations interested in pursuing follow-on contracts.

IMPLEMENTATION

Activity 9A: Prepare Public Relations Material

Good public relations may be achieved if an open house or some kind of ceremony occurs to mark the formal delivery of the completed project.

Good public relations may be achieved if an open house or some kind of ceremony occurs to mark the formal delivery of the completed project. Speeches, ribbon cuttings, buffets, or even banquets are an opportunity to focus public attention on the completed project and to thank the workers, managers, and funding sources involved in project development. Photo opportunities may also be planned so a visual record of delivery and acceptance exists.

PROJECT EXAMPLE

An open house at the activity center is held to mark the formal transfer of the facility. Town officials, representatives from the funding sources, and supportive Northeastern faculty attend. A brief ceremony occurs where responsibility for the activity center is transferred from the team members to the town of Bedford. The document that transfers responsibility to the town is signed, and Roger delivers a box to Mr. Williams that contains these items:

- The warranties, appropriate invoices, and instruction manuals
- The most recent copy of the Production Specification
- A complete set of corrected current Working and Detail Drawings

A buffet—paid for out of the university's public relations budget—is provided for everyone to enjoy.

The students request Mr. William D'Annolfo to step forward. They present to him a certificate of appreciation that they have secretly prepared for "Uncle Bill." It expresses their thanks for his continuing advice, support, and encouragement during the entire project. He thanks them and reminds them that, thirty years ago, he was a student at Northeastern, majoring in chemistry and civil engineering. He tells them that, with persistence, good work, and a little luck, they can follow in his footsteps. He also informs them that he has one opening for a co-op student starting the following January. They all excitedly discuss who should apply to Uncle Bill's company for the assignment.

Activity 9B: Offer Presentations to Prospective Customers

Prior to completion of any current project, new customers are needed. Individuals who had been active in the Study and Design Phases and whose abilities could apply to similar projects can be transferred to a sales and marketing group so they can pursue new business opportunities together. Presentations are planned, rehearsed, and given to potential customers. These customers may sometimes want to examine previously completed projects. Permission should be requested from previous customers for site visits by the potential new customers.

Activity 9C: Construct Speculative Products

In some instances a contractor will build a structure without having an identified buyer. This is known as **building on speculation.** It means that

- The builder's business "contracts" with itself
- The builder defines what is to be built
- The builder obtains the financing

For example, software programs can be devised by individuals, groups, or companies on speculation. These programs can be demonstrated at conventions or seminars. Mailing lists may be purchased, and advertising literature prepared and sent to potential customers. Follow-up calls are then made to determine if potential customers are interested in purchasing the software. For more complicated software programs, demonstrations of software at a potential customer's location may be planned and made via a so-called road show.

SECTION 5.3   IMPLEMENTATION PHASE DOCUMENTS

During the Implementation Phase, implementation or construction proceeds with the Production Specification and Working and Detail Drawings as a guide. Implementation, of course, never perfectly proceeds according to what is described in the above documents. Thus, Contract Change Notices (CCNs) and Engineering Change Notices (ECNs) are prepared, as approvals to modify the design or design approach are obtained. An example of a change is given in appendix E.

The content of each of these documents is described below. The flow from the Conception Phase documents to the Implementation Phase documents is shown in figure 5.7.

Although the project documents from the Design Phase define the project, implementation, of course, never perfectly proceeds according to what is described in them. Thus, Contract Change Notices (CCNs) and Engineering Change Notices (ECNs) are prepared, as approvals to modify the design are obtained.

As implementation progresses, the approved Contract Change Notices are gathered. The Working and Detail Drawings are modified to agree with these design changes. Instruction manuals are written to explain how to operate and maintain the equipment. More meetings are held to evaluate progress; more meeting reports (section 8.3) are written. These reports and modified drawings eventually lead to three Implementation Phase documents. At the end of the Implementation Phase, the completed documents are the following:

1. Instruction Manuals and Warranties
2. Maintenance Specifications
3. As-Built Plans

The content of each of these documents is described below:

Here is how to prepare the **Instruction Manuals** and **Warranties:**

During the Implementation Phase, documents need to be prepared to deliver to the client. Most important are instruction manuals, which describe how to use the product.

1.   Promptly after receipt of goods, instruction manuals must be obtained for each piece of equipment that is a part of the implementation. For the other portions of the implementation, typically those that the team developed as project-specific, the team must write its own system-oriented instruction manuals. These manuals might reference the manuals supplied by your vendors, including page and paragraph number. When a simpler set of instruction manuals is desired and all information is being gathered into one document, those parts of vendors' manuals that apply to the particular equipment or system can be extracted.

2.   Promptly after receipt of goods, warranties from vendors must be obtained, and warranties must be written for equipment the team developed. For equipment supplied by vendors, the team must provide to the client copies of the applicable vendor invoices that show proof of purchase. Thus, if an item that the project team purchased fails, it can be repaired or replaced at a client's location by the original source or manufacturer—usually at *no* cost to the project team's firm.

The **Maintenance Specifications** are prepared for large systems during the Implementation Phase. For small systems, maintenance information is delivered to the user in the form of one or more manuals.

1.   For large systems, a study is performed near the end of the Design Phase. It identifies potential single failures and potential multiple failures caused by the first (single) failure. This study then leads to the outline and preparation of one or more interrelated Maintenance Specifications. These specifications are often divided into routine preventive maintenance, which can be scheduled, and random maintenance whose need cannot be predicted in advance. (See appendix G.)

2.   For small systems, copies of vendor-supplied maintenance and troubleshooting manuals are provided to the user. The designers may prepare an overview document that guides the user to the proper vendor-supplied manuals and may also include interface information regarding

IMPLEMENTATION

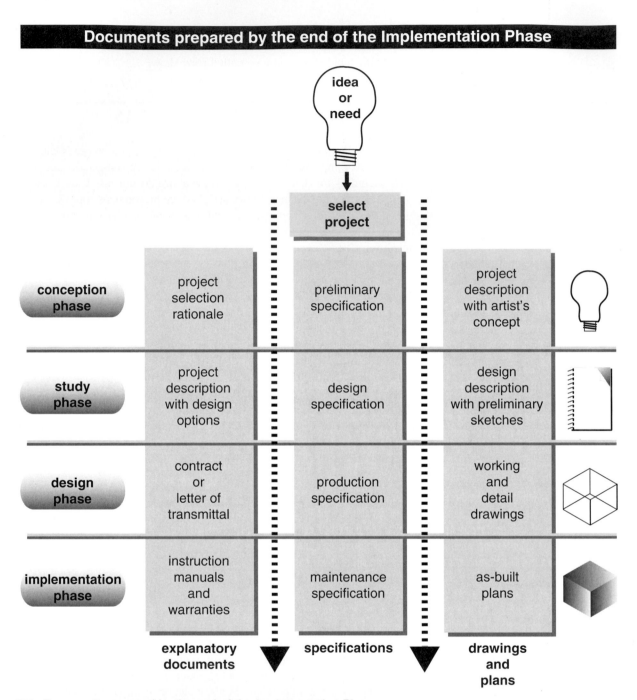

Figure 5.7   Documents prepared by the end of the Implementation Phase

routine maintenance and random maintenance. This additional documentation ties together the small system parts.

The **As-Built Plans** are prepared as follows:

    1.   All change notices must be examined and these revisions included in the documentation being prepared for delivery to the client and/or funding source(s). All construction or manufacturing drawings containing changes that were not properly documented with change notices should be determined; descriptions of each change need to be inserted into their proper place in the overall documentation. For large projects or programs, it may be necessary to issue change notices for any undocumented changes. (People who are busy working often forget to document their changes.)

2.   When a package of the corrected working and detail drawings has been gathered together, it must be accompanied with written material that describes the changes that have occurred during construction (or manufacturing). The written material must explain why these changes occurred. Thus, both the client and the funding source(s) will have documentation that is correct and up-to-date and that also includes explanations as to why the changes occurred during the Implementation Phase. Appendix H lists the content of the Student Activity Center As-Built Plans.

The As-Built Plans describe the project as it was actually completed, with all revisions to plans included. These plans and accompanying records are valuable to the client as help in using the product and to the group that performed the project.

Upon completion of a project, a final report should be prepared. It should contain a description of the project as it evolved and progressed. It should also include information regarding how the final design became the specific product that was implemented. The physical and electronic location of all related data and documents should be noted if it becomes necessary to refer to them at a later time.

SECTION 5.4   DEFINING PHASE COMPLETION

The Implementation Phase is considered complete when all of the following are accomplished:

- The documentation of the project implementation is complete.
- The customer has accepted the project.
- Final payment has been received for the project.

Now the project team can be disbanded. The team members can go on to a new project.

SECTION 5.5   ACCOMPLISHMENTS OF THE SYSTEMATIC APPROACH

As a review of chapters 2 through 5, a systems approach to a project requires that you accomplish the following:

- Formulate the problem versus its symptoms, recognizing the real versus perceived scope of the problem (Conception Phase)
- Analyze the data (Conception Phase)
- Search for alternatives (Conception Phase)
- Evaluate alternatives (Study Phase)
- Devise a solution (Design Phase)
- Implement the solution (Implementation Phase)
- Review the results (Implementation Phase)

Projects that operate by methodically applying a systematic approach will more likely achieve all three of the following desired outcomes:

- Complete all assigned tasks
- Remain on schedule
- Complete the project within budget

Projects are more successful when they are approached and performed in a systematic manner.

CHAPTER OBJECTIVES SUMMARY

Now that you have finished this chapter, you should be able to:

1. Define the purpose and goal of the Implementation Phase of a project.
2. Explain the tasks involved in the Implementation Phase of a project and the activities or steps through which these tasks are performed.

IMPLEMENTATION

3. Describe ways in which money needed for projects is typically obtained and delivered.

4. Explain how documents prepared during the Design Phase are used in the Implementation Phase.

5. Describe tasks that are typically part of the Implementation Phase and how they are assigned.

6. Describe the role of the project manager during the Implementation Phase.

7. Explain why periodic meetings are important during the Implementation Phase of a project.

8. List common tasks that should be accomplished before actual implementation or manufacturing begins.

9. Describe what prototypes are and why they are developed.

10. Define the learning curve, and explain how it relates to manufacturing production.

11. Describe the documents that should be prepared for the Implementation Phase and which should be delivered to the client.

12. Define "delivery" of a project, and describe what problems may occur at that stage.

13. Outline the responsibilities and opportunities for a supplier/manufacturer after the delivery of a product.

14. Explain why a review meeting at the end of a project is important.

EXERCISES

5.1   Revise the Purpose and Goal statement given in section 5.1 to match each of your options for a class project.

5.2   Carefully examine the activities listed and described in section 5.2. Adapt them to each of your options for a class project.

5.3   Review the Design Phase effort with all team members. Develop any required additional Implementation Phase details with the assistance of all team members. Organize the team into task groups for the Implementation Phase, supported by personnel whose Implementation Phase skills are necessary. Assign each group a set of Implementation tasks to perform.

5.4   Develop an applicable CCN form and procedure for your project.

5.5   Contact selected vendors about exact schedule and delivery dates. Verify that all federal, state, and local requirements are being achieved. Sign contracts with all involved parties.

5.6   Document the Implementation Phase changes. Gather together and deliver warranties and applicable invoice copies. Review the results of the Implementation Phase.

5.7   Formally deliver the final product with its appropriate manuals and warranties. Terminate the Implementation Phase.

5.8   Describe how you would supervise contractor workers versus how you would supervise your other team members.

5.9   Describe how you would determine the optimum time to terminate the Implementation Phase.

5.10   Write a final report for your completed project.

5.11   Write a paper describing what you have learned from your project. Indicate how your effort has improved your technical and personal skills and how this effort has affected your choice of a professional career, if applicable. Revise your resume to reflect your newly acquired skills and experience and, if applicable, a change in your career direction.

chapter **6**

1. What are the four key responsibilities of managers?

2. How do the role and responsibilities of a manager change over the phases of a project?

3. What is a typical organizational structure? What is the role of each group in the organization in project planning, performance, and implementation?

4. What are the two main styles of management, and how do they differ? What are the advantages and disadvantages of each?

5. How much should managers consider individual skills and interests in choosing people to assume certain roles in a project?

6. Why are performance reviews important, and what are the various ways in which people can be evaluated?

7. What are spreadsheets, and how can they be used by managers?

8. What are common databases, and how can they be used by businesses?

9. What are the responsibilities of project and program managers during the performance of a project or program? What is their role in controlling and reviewing the project or program?

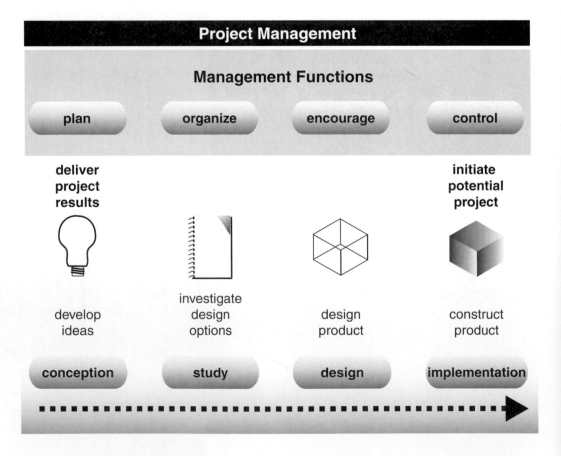

Nothing is impossible if you do not have to do it yourself.
Adapted from an A. H. Weiler New York Times *memo*

Individuals who possess a combination of management skills and technical knowledge are vital to the growth of business and industry, particularly for those organizations involved in highly technical fields.

This chapter focuses on the skills, requirements, and responsibilities associated with managing a project. The skills required to manage projects, programs, and people, combined with technical knowledge, are in great demand. Individuals who possess a combination of management skills and technical knowledge are vital to the growth of business and industry, particularly those organizations involved in highly technical fields. Much has been written on this subject (Archibald, 1992; Badiru, 1991; Badiru and Whitehouse, 1989; Chase, 1974; Farkas, 1970; Francks, Testa, and Winegardner, 1992; Lewis, 1991; Samaris and Czerwinski, 1971).

After you complete the study of this chapter, you should be able to accomplish the following:

- Identify management functions, organizations, and styles.
- Select people to staff a project.
- Read a spreadsheet.
- Gather and organize information necessary to monitor and control projects.
- Understand the components of project and program control.

Duncan (1996) observes that both leading and managing are required; one without the other is likely to produce poor results. Managing is primarily concerned with "consistently producing key results expected by stakeholders." That is, a successful manager contributes to the financial well-being of the organization. Leading, on the other hand, involves establishing direction, aligning, motivating and inspiring people—helping people to energize themselves to overcome barriers to change. Success requires the participation of all members of the team, but it remains the responsibility of managers to provide the resources needed for the team to succeed.

There are several important questions regarding management functions, organization, and styles. Each question will be presented below, with our opinion as to the answers. For each question, the general answer will be presented first, followed by a specific example from the activity center project (described in chapters 2 through 5) where this is applicable.

SECTION 6.1   MANAGEMENT FUNCTIONS

What is management, and what is the goal of management? These questions are discussed in this section.

*The group of people known as **managers** must plan projects, organize them, encourage workers, and control projects to ensure quality performance within the allotted time and budget.*

What Is Management?   Management is the name given to the role of guiding and coordinating the activities of a project team. The group of people known as **managers** must plan projects, organize them, encourage workers, and control projects to ensure quality performance within the allotted time and budget. Here are some details of the four key management functions—plan, organize, encourage, and control:

Plan: Managers must ensure that a project moves smoothly and in its proper direction by planning the work to be accomplished and selecting the appropriate personnel to perform the work.

Organize: Managers must establish or enforce the use of standards for reporting and record-keeping in order to approve the purchase of materials, monitor contract performance, and control expenditures. Managers in an efficient organization need to examine reports and other records without expending a great amount of time and energy interpreting their content. Therefore, they need reports and other records presented in standard formats. If the formats are already established, managers merely have to enforce their use; but if these standards do not exist, the managers must develop them.

The manager of a well-organized project knows that written documentation is important. It is used to judge the status of a project as it develops. Managers must continually examine when to end a phase and when to initiate the next phase. They need to ask questions such as this: How much time should be allowed for project conception, study, design, and implementation? The longer the time a project requires—from conception through delivery—the higher the costs. Yet enough time must be allowed to correctly accomplish the requirements of each phase.

Managers must participate in the development of contractual documents because the final responsibility for fulfillment of the contract provisions falls upon them. Managers must also recognize their lack of knowledge in certain areas and so obtain the consulting services of specialists. Larger projects and programs often include an attorney and one or more technical advisors on their staff to help with legal and contractual issues.

It is the responsibility of managers to encourage team behavior on the part of their employees. They also need to promote training and growth on the part of employees.

Encourage: Employees work best when they work well together, and they need to feel that they are a part of the team. It is the responsibility of managers to encourage employees to perform as a team. Special courses are offered to technical managers to assist them in becoming more sensitive to the needs of employees and to guide them in a positive manner. Notice the use of the word *guide* rather than the word *direct*. As will be discussed in section 6.3, guiding workers is a more effective approach than directing workers. Brown (1992) offers the following suggestions for positive guidance:

- Treat people fairly and honestly.
- Give credit to those who assist you in any way.
- Be consistent—that is, let people know what to expect from you.
- Act in ways that avoid personal loss or embarrassment.
- Be a good listener.

It is generally agreed that the training and growth of employees has a positive effect on the growth of a company. However, there is a continuing debate in the field of management training regarding *how, when,* and *where* to encourage the growth potential of their employees. Some texts state that excellent technical specialists should be offered promotions that encourage them to increase their technical abilities. Other texts state that excellent technical persons can be trained to become good managers. The *how* can be accomplished by offering employees both technical and management courses. The *when* can occur during a formal performance review or during a meeting when the employee is reporting upon work status. The *where* may be company-sponsored training sessions at the office or company-paid courses at local colleges and universities.

Control: Managers monitor and control contract performance (tasks, schedules, and budgets) in order to see that the projects assigned to them are completed on time. They must also provide a quality product, perform within cost, and see that the people involved continue to work well and efficiently together.

Managers control expenditures, which include payments to subcontractors, project workers, and material suppliers. The manager of a project must know if actual expenditures exceed the estimated expenses and the effect of any budget overruns on the ability to complete the project. They must be able to judge the willingness and the ability of the lending or sponsoring financial institution to cover such additional expenses. The manager must continually compare the project expenditures with those funds allotted for the project accomplishment.

Managers approve the purchase of materials in order to ensure that any significant material purchase is at a price that agrees with the original estimates. Signature authority can be used to control these expenditures. It is necessary to decide when to insist on additional signatures and when that requirement leads to slowing down the project's progress. The decision as to the level of signature authority required may change as the project progresses from one phase to another.

The manager must become a skilled negotiator, as compromises are often necessary to achieve the overall project goals. In such situations, the manager may be acting more like a politician.

In order to guide the progress of a project, managers must have access to accurate and timely information, particularly regarding financial matters. The technology of today provides the capability for automated computation and revision of financial documents known as spreadsheets—the topic of section 6.5. Managers must verify the performance of previously authorized or approved labor, and they must obtain assurance that the material received from suppliers agrees with the description on the purchase order; that is, the received material is correct in quantity, cost, and quality. The manager must become a skilled negotiator, as compromises are often necessary to achieve the overall project goals. In such situations, the manager may be acting more like a politician.

What Is the Goal of Management?

The goal of management is to obtain the optimum quality output from individuals and groups by performing the four key management functions for the least investment of time, money, and other resources.

The emphasis placed on each of the four key management functions—plan, organize, encourage, control—shifts during the various phases of a project.

All four key management functions are important. However, the emphasis given to each of these functions may change from one phase of the project to another.

1.   The ideas created and the work accomplished during the Conception Phase will set the tone of the entire project. Thus, the manager must encourage all involved workers to be as creative as possible. Planning the remainder of this phase and the tasks, schedules, and budgets for the remaining three phases is also important but will be meaningless unless one or more creative solutions for the project are devised.

2.   The plans that evolve during the Study Phase will directly affect the direction and cost of the remainder of the project. Additional human and material resources are being applied at this time; more will be required later. Planning now becomes important so all tasks are accomplished and the project stays within budget and schedule. Encouragement is still important because new ideas often evolve during the Study Phase. Organizing for the next phase is growing in importance.

3.   The Design Phase must proceed as organized. A plan must be prepared for the Implementation Phase that is the most effective plan for the proposed design, even while the design details are being formulated. Control assumes greater importance because the cost of redesign will cause the cost of the project to increase greatly. Most designers need some supervision and encouragement so they will choose the optimum design details while not losing track of the goal of the entire project.

4.   Control is vital during the Implementation Phase. All theoretical work is now being converted to actual practice. This phase seldom proceeds smoothly without careful control of the tasks, schedules, and budgets being implemented. Workers must be encouraged to work within the plan or else promptly report any needed revisions to the manager. The manager must continually collect data and compare the results of the project work with the expected performance. When a difference between the planned or expected performance and the actual performance is detected, the manager must make the decision to modify the pattern of performance by adding or removing resources such as people, money, or equipment. Organizing now becomes reorganizing and should occur only when necessary. Additional planning should seldom occur except when a detail of the original plan proves to be impractical.

The following explanation shows how the key management functions were implemented for the activity center project, the example in chapters 2 through 5.

During the Conception Phase, Professor Hulbert encouraged the team and guided the initial planning. Organizing the team into small groups and controlling the time available were less important. The team members gradually took over the role of managing the project, with Professor Hulbert taking more of an advisory role as he realized that the team approach was operating successfully.

During the Study Phase, Professor Hulbert first emphasized planning, followed by encouraging the team members to focus upon potential designs. During the Design Phase, he tried to ensure that teams were effectively organized into groups capable of determining design details.

During the Implementation Phase, Professor Hulbert stressed the need to control the schedule and monitor costs carefully so the team would stay within the allotted time and budget. Much of the actual management was taken over by Roger, the project team leader. Later Roger was assisted by Alice, who oversaw the technical aspects of the activity center project.

Managing involves both people and paperwork. People prepare tasks, schedules, and budgets. Some people perform, other people evaluate, and other people ultimately accept the results—the final product. Much coordination is required, even for a project as small as the activity center.

SECTION 6.2   PROJECT ORGANIZATIONS

This section addresses the following three questions:

- How can people be organized to be productive and efficient?
- How are large projects and programs organized?
- How is a small project organized?

Organizations help managers fulfill the functions described in section 6.1. Organizations assist managers in handling personnel and organizing projects.

How Can People Be Organized to Be Productive and Efficient?   Any organization
that expects to survive and grow after all its existing projects and programs are completed must consider the issue of preparing personnel for advancement. This is an important part of encouraging the personnel, who then will usually act as good team members and become both more productive and more efficient.

For the organizational chart example shown in figure 6.1, Sales and Marketing (staff) advises the president and executive staff regarding customers and competition. The Program Office (staff) watches over programs being performed by the Research, Engineering, Manufacturing, and Quality Control (line) operations. Personnel and Purchasing (support) assist both the staff and line groups.

A firm can organize each one of their functional areas by product line, customer type, or some other type of grouping. For example, one major manufacturing firm is organized according

*In many corporations, employees are organized into groupings known as **staff, line,** and **support.** These categories describe whether the groups advise (staff), direct and perform (line), or provide support activities for other groups.*

Figure 6.1 Typical organization chart showing staff, line, and support functions

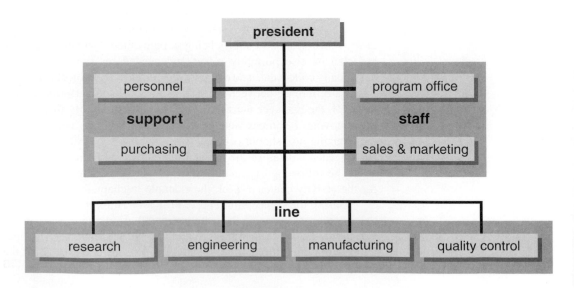

Typical organization chart showing staff, line, and support functions

to type of product. It has a commercial division and a military or government division, each one of which has its own functional divisions (i.e., manufacturing, sales, marketing).

Projects can emerge from a firm's organizational structure or they may result from the good idea of a manager, worker, or consultant. Regardless of who was responsible for the initiation of a project, the project must be organized to encourage creativity and to promote the achievement of the goals and objectives of the firm. The next section discusses a range of organizational forms that are used for organizing projects and relating them to the parent organization.

How Are Large Projects and Programs Organized?   The organization structure adopted for a particular project will depend upon the firm, the project goals, and the size of the project. Generally, it is recognized that there are three basic organizational forms for projects. Most projects are organized according to some variation of the standard organizational structures.

Stand-Alone Project Organization   This type of project organization is sometimes referred to as a pure project structure. In the stand-alone organization, the project is organized similar to a small company. Figure 6.2 illustrates a program that consists of two projects organized according to the stand-alone configuration. In this organization structure, the project manager has full and total control of the project. Usually the project has a complete staff of workers. Therefore, it is unnecessary to "borrow" individuals from other projects or from the parent organization to complete the work. In small corporate organizations, the project manager for a stand-alone project will report directly to the highest level of management. Even in a large corporate environment, the manager of a stand-alone project will report at the vice president level. The organization of a stand-alone project is usually easy to understand—very flexible with decentralized decision making, and a strong sense of team spirit.

Some of the features that are considered to be benefits of the stand-alone organizational structure can become difficult to work within at some point in the life of a project. For example, the sense of team spirit is good at the beginning of a project, but can be a problem at the end of

Figure 6.2 Stand-alone project organization

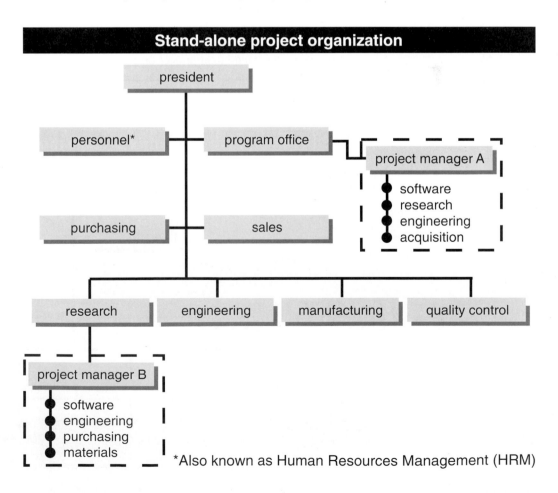

a project's life. Why? The members of the team become so involved that they do not want to see the project end. Also, sometimes stand-alone projects result in overstaffing. The utilization of equipment that is already available in the parent firm is avoided; the project personnel "want" their own equipment.

Stand-alone projects are often focused on research with the intent of developing something new for the firm, such as a new product or a new production method. This project organizational structure is used by many large firms for writing proposals to potential customers. As an example, a major defense contractor rented a vacant building that had been a supermarket and assembled a team of several specialists to write a proposal to build a weapons system. The project was performed entirely "off site" (away from) the parent firm's buildings.

The Matrix Organization   Unlike the stand-alone project, the matrix organization is highly dependent upon the parent organization. The matrix organizational structure is a very popular form of configuration for firms wih many projects that have very similar labor requirements. Figure 6.3 illustrates the matrix form of project organization. Generally, in order to use this form of project organization effectively, there must be a project manager who is responsible for organizing and assisting in the implementation of the individual projects. Each project has its own project manager. The parent firm is organized in such a way that functional areas such as manufacturing, research, and engineering stand alone and provide a "home" for staff members who have the respective areas of expertise. The managers of the individual projects are provided with a budget to complete the work. It is the job of the project manager to negotiate with each functional area manager and obtain the type of expertise needed to complete the tasks on their project. For example, a project manager may estimate that she needs one thousand hours of software engineering work to complete the software development aspects of her project. The project manager would go to the Software Engineering Department manager and negotiate for one thousand hours of work from the functional manager. For each of the tasks a project manager needs completed, she will "buy time" (purchase services) from the appropriate functional group. At some point in time, it is possible in the matrix structure for the project manager to be the only full-time person working on the project.

Similar to the stand-alone project organization, the matrix organization has one individual who assumes the responsibility for completing the project on time and within budget, while

Figure 6.3 Matrix organization

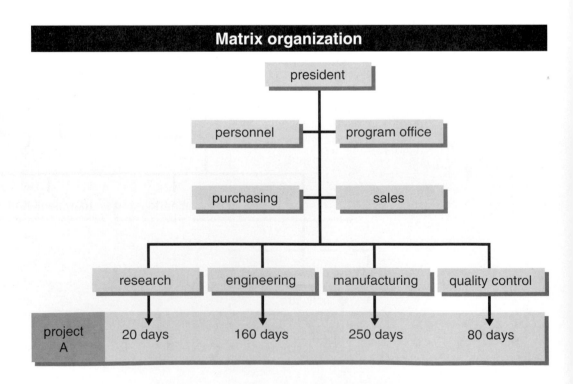

satisfying the customer requirements. The matrix organization has access to as many specialists and areas of expertise as the parent organization. From an overall corporate view, the matrix-management approach to managing projects allows the firm to realistically plan for and organize its resources.

There are some drawbacks in matrix-managed projects. The project manager is required to continually negotiate with functional managers in order to obtain the required personnel for a project. Individuals are not always available at exactly the time a project needs them. Scheduling is a major challenge in most matrix-managed projects. Project managers are sometimes competing for the same people, resulting in time-consuming politics and competition among the project managers.

The Embedded Project Organization   In some large organizations, it is common practice to integrate (embed) projects into the functional departments of the firm. For example, a project that involves some research may be assigned to the Vice President of Research and Development for completion. A project that has a focus on developing variations in existing products can be assigned to the Design Department. Projects organized according to their function are generally "picked up" (acquired) by the group that can provide the highest probability of successful completion. Generally, projects are embedded in the line functions. Figure 6.1 illustrates a typical organization chart.

Projects that are embedded in a functional organization can very easily draw upon the technical expertise of individuals employed within the organization. The biggest drawback to having a project embedded in a functional division is that corporate-assigned work generally takes precedence over the project tasks. Response to the client's needs may be less than enthusiastic within the organization. Most embedded projects are not in the mainstream of daily activity. On the other hand, when a project is very important to the corporation, embedding it in a functional division will ease the scheduling of necessary resources to get the work done on time. People with varying levels of expertise can focus on completing the work according to the customer's requirements.

Project Organization in Practice   It is a very rare instance in which a project is organized exactly as a stand-alone project, a matrix-organized project, or an embedded project. Most projects follow a mixed organizational format. Usually, well organized project managers adopt the features that promise successful completion of the project. For example, the manager of a project that requires substantial software development effort may employ ("bring on board") several software engineers for the life of the project. From the software development perspective, the project may appear to be organized as a stand-alone project. Other functions in the same project—such as manual writing and testing—may be performed by individuals whose time is negotiated with their functional manager. From the perspective of the manual-writing function, the project appears to be organized in a matrix format.

The president of a company must organize employees into efficient and effective groups and encourage the group managers to explore new paths for the corporation. The president must focus on the long-term goals of the corporation. The president should insist first on planning and must continually review plans to verify that they are either being followed or are being revised. Control is exercised only when a group does not meet its goals. This often leads to reorganizing, replacement of managers, or both. This management technique is known as **managing by exception.**

The president of a company and executive staff must focus on the long-term goals of the corporation and prepare long-range plans, as well as monitor whether current goals are being met.

Sales and Marketing personnel are responsible for moving the products into the marketplace, monitoring the marketplace, and investigating and planning new directions for the corporation. They must also encourage the line groups—particularly Research and Engineering—to investigate new technologies that will improve the corporation's position in the marketplace.

The Program Office personnel must plan, control, and verify the status of large projects and programs. They often support Sales and Marketing in the pursuit of new business opportunities.

The Personnel group, often known as Human Resources, retains employee records, including periodic evaluation of all employees. They can help employees no longer required on a project to find other work. They often provide employee guidance and encouragement in developing new skills via either in-plant seminars or local college and university courses and programs.

The Purchasing group buys needed materials, equipment, and supplies. Purchasing personnel should combine the purchase requisitions for common items needed across departments into a single purchase order. They can then negotiate the best price and delivery with suppliers and verify the receipt of the ordered material. They often monitor the Engineering and Manufacturing line groups to encourage the use of common items.

The line groups of Research, Engineering, Manufacturing, and Quality Control are responsible for the performance of day-to-day activities related to company projects. (Quality Control should report to the president to ensure top management attention.) Line managers organize, plan, encourage, and control their workers as the workers perform their assigned tasks.

Other support groups not shown on our simplified organization chart include Shipping and Receiving, Accounting, and the president's executive staff.

How Is a Small Project Organized?   The activity center project described in chapters 2 through 5 was a school project. It provides a good example of how a small project can be organized. See figure 6.4. The team leader reports to a teacher or dean who is responsible for encouraging, monitoring, and controlling the project as it develops. The teacher or dean is equivalent to a project office manager, whose role is to assure the school administrators that the project is progressing satisfactorily. The specialty consultants and construction company personnel support the involved student workers in the performance of the project. Such an organization is referred to as **project oriented.**

The activity center project team is organized as indicated in figure 6.5. The project team leader (Roger) has three types of groups reporting to him:

- Specialty consultants
- Student workers
- Construction companies

The specialty consultants include the following:

- Professional people who are registered to perform their profession within the state in which the work is to occur
- Tradespeople who are licensed with the state for their particular trade

Figure 6.4 Organizational chart for the activity center project

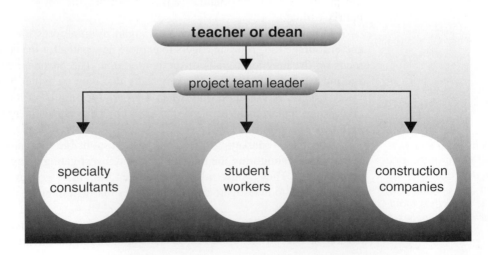

Organizational chart for the activity center project

teacher or dean

project team leader

specialty consultants

student workers

construction companies

Figure 6.5 Organizational details for the activity center project

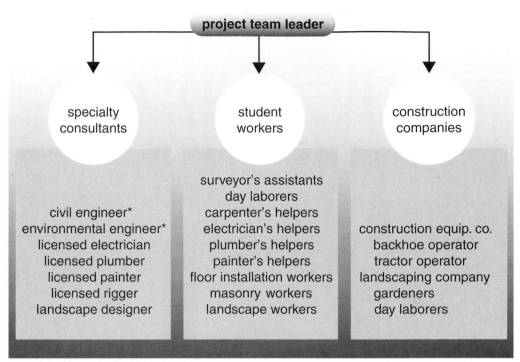

Organizational details for the activity center project

project team leader

specialty consultants

student workers

construction companies

civil engineer*
environmental engineer*
licensed electrician
licensed plumber
licensed painter
licensed rigger
landscape designer

surveyor's assistants
day laborers
carpenter's helpers
electrician's helpers
plumber's helpers
painter's helpers
floor installation workers
masonry workers
landscape workers

construction equip. co.
backhoe operator
tractor operator
landscaping company
gardeners
day laborers

*registered professional

They agree, via a contract, to assist the activity center team in those areas where their special skills and registrations or licenses are required.

The student workers are those personnel whose skills are required to perform the work scheduled. Most students are flexible enough to perform a variety of duties under the supervision of the professional people or tradespeople.

In this example, the construction companies consist of two separate organizations:

- The construction equipment company that provides

    Expertise in construction management

    Construction equipment

    Skilled workers who are capable of operating the construction equipment

- The landscaping company that provides

    Landscaping equipment

    Landscaping skills

    Skilled workers who are capable of implementing the landscaping design in a satisfactory manner

These companies are flexible enough to allow the students to perform many of the laborer duties under their supervision.

For the activity center project, the workers, laborers, and helpers are supplied from the student team wherever state and local laws allow. In this manner, students participating in the project gain experience in design, development, and construction.

SECTION 6.3   MANAGEMENT STYLES

How Do Managers Manage?   There are two different styles of management: **top-down management** and **bottom-up management.** Their advantages and disadvantages are described in this section. Most organizations apply a combination of these two styles.

In top-down management, a leader directs and gives orders; the result can be quick response to problems. In bottom-up management, all levels of workers participate in decision making; the result can be that all workers feel involved in the project.

1.   Top-down management is an organizational structure where the highest-ranking leader is personally involved in directing all other personnel. The leader directs through verbal orders and written directives. The leader does not expect or want to be challenged regarding decisions. This management approach can result in quick responses to information and to problems that occur. It is very effective on small projects where the workers report directly to the manager. However, on large projects, the leader may not be able to provide the level of attention needed and may lack enough detail to be efficient or effective. Workers' morale may often be poor because the workers do not feel involved.

2.   Bottom-up management is an organizational structure where the lowest level of personnel offers suggestions that are gathered by those individuals on the next level of hierarchy. This next level of managers examines the ideas of the workers, coordinates these ideas with other managers, inserts ideas of their own, and recommends these coordinated ideas to the next level of managers. This management approach is slow in response to information available. However, it may be more efficient and effective in the long run. Because the workers feel involved, morale is usually high and, as Brown indicates (1992), good human relations are crucial to change and improvement:

- Getting good information from people requires their cooperation
- Good ideas come from people who are properly motivated
- Receptivity to new ideas is more easily generated when people feel part of the decision-making process

For any style of management, managers must be excellent listeners. Managers must gather, absorb, and disseminate information in order to contribute to and control a project.

For any style of management, managers must be excellent listeners. Managers must gather, absorb, and disseminate information in order to contribute to and control a project. Managers need to work to reduce or eliminate dissension, strife, and perhaps even hostility among team members. Managers need to analyze significant differences of opinion to avoid alienating individuals and offer viable solution options. This technique is known as **conflict resolution.**

Conflicts may occur over the following issues:

- Specifications
- Schedules
- Budgets
- Tasks
- Activity priorities
- Resource allocation
- Goals
- Expectations
- Technical approach
- Personnel availability
- Personality differences
- Administrative procedures

Archibald (1992, p. 99) discusses the types of conflicts that must be resolved by the project manager:

1. Conflicts over project priorities
2. Conflicts over administrative procedures

3. Conflicts over technical opinions and performance trade-offs

4. Conflicts over people resources

5. Conflicts over costs

6. Conflicts over schedules

7. Personality conflicts

There may also be an issue regarding how much responsibility is assigned to the project manager, who may not have the authority to resolve a conflict. Such a problem must be resolved with those to whom the project manager reports.

Project managers can become diverted from their technical tasks when they must focus upon the supervision of people. Project managers must learn to continually weigh how much time they can invest in the technical details of a project and how much time they must allot to personnel problems.

One central role of the manager is to effectively resolve any conflicts that occur during the implementation of a project.

Managers can ignore the conflict for a while and see if it resolves itself. They can also force a resolution by insisting on a particular solution. One way of resolving conflicts that involves the team members is to perform the following actions:

• Request memos describing each point of view

• Convene a meeting

• Obtain an agreement for an acceptable compromise

This allows all involved some degree of recognition of their ideas.

During the conflict-resolution process, the manager must continually remind individuals that everyone is a member of the same team. The overall goal must be to effectively use the project's resources to complete the project on time and within the allocated budget. Another goal is to create an environment that results in all persons feeling like they have "won." (The socalled win-win situation is the best result possible.)

It is important to remember that there would be no conflict if the people involved did not care about the project. Conflicts can have their benefits if they are handled properly. As a result of successful conflict mediation, there may be a greater understanding of the project by everyone involved.

The activity center project was managed using primarily a bottom-up approach. Most of the time Roger discussed with the other students, in advance, the options for the project that he and Professor Hulbert had worked through together. Sometimes Roger would announce how the work should now proceed.

Note that the team was usually given the opportunity to react to Roger's decisions. In this manner, the team members felt involved and could better perceive the reasoning behind all major decisions. Team members also had a better view of how their work related to the work of others and to the overall project. This primarily bottom-up approach leads to a more efficient and effective result, except where personality clashes occur.

The effort of avoiding clashes and the large amount of responsibility given to one individual are reasons why many technically oriented persons prefer to avoid management responsibilities and concentrate on the technical effort. However, effective handling of conflicts is a key responsibility of managers.

SECTION 6.4   PROJECT STAFFING

This section focuses on how to staff a project for the most effective use of the available personnel and how the manager controls the assignment of staff members. Two questions are discussed:

• How are people selected to perform the work involved?

• How and when should people be evaluated?

Each of the four phases of a project needs a different combination or "mix" of skills on the part of the staff. The needs for creators, planners, performers, laborers, and administrators are also described in this section. Conflicts and their resolution are discussed in detail.

Managers are responsible for locating properly skilled personnel who are capable of assisting in the implementation of their project. Sometimes existing employees recommend friends or former colleagues. More often, a manager must either advertise for the personnel needed or use an employment agency. First the manager prepares a job description detailing the responsibilities and related skills needed. The job description may indicate the level of management to whom the future employee will report and the type of people the future employee may supervise. Resumes of potential employees' skills and interests are sent to the manager for review. Those people who appear most qualified based on their resumes are called in for a face-to-face interview. Based on these interviews, the manager makes a final selection.

In the assignment of tasks, the work preferences of individual personnel should be considered. However, to counterbalance this desirable goal, costs and schedules on a project must be controlled, and the required work must be performed and completed properly. All those working within an organization or working on a project must perform as members of a team rather than as individuals concerned with only their part in the effort. From time to time, the mix of skills of individuals within the organization will change; revising this mix will need to be managed effectively and on a continuing basis (Connor and Lake, 1994).

*Managers must be able to assess which people are best at what types of tasks and sense people whose skills can develop while working on the project. The manager must also be able to recognize how people will work together. These managerial capabilities are known as **people skills.***

The manager must be able to sense which people are best at what types of tasks and assess people whose skills are most likely to develop as the project planning, performance, and control progresses. The manager must also be able to recognize how personalities interact, and thus assign people so that they interact in a positive and constructive manner. These managerial capabilities are known as **people skills.** The development of the team starts with the manager and that manager's organizational abilities. (The issue of the role of the manager in resolving conflict was discussed in section 6.3.)

People have different skills and interests. They may be creators, planners, performers, laborers, or administrators. Creators and planners are essential in the early phases of projects; performers, laborers, and administrators are instrumental in the later phases of projects.

How Are People Selected to Perform the Work Involved?

Each person has certain skills, capabilities, and interests. One person may like to build, another may prefer to plan, others like to negotiate, some enjoy organizing, and others prefer to supervise. People are selected based on how their training, skills, and interests fit with the needs of the project and which phase of the project is in progress—as noted in chapters 2 through 5.

Individuals may be grouped into a variety of categories according to skills and capabilities. Here are some of these categories:

* Creators
* Planners
* Performers
* Laborers
* Administrators

Each phase of a project requires a different combination of categories (see figure 6.6).

Creators: For the Conception Phase of a project, the necessary skills are those of imagination and vision. Creators or conceivers are people who can think about a problem and envision one or more solutions. Sometimes these people are also capable of developing the details of a concept. Other times, such people must transfer their ideas to the performers.

Planners: For the Study Phase, the most desirable of the proposed projects are to be developed into potential design projects. Planners are now very important to the eventual success of a project. Planners, such as cost estimators, like to work with numbers and analyze them. They enjoy planning the details of a proposed project and preparing and analyzing estimates.

Performers: For the Design and Implementation Phases, persons who enjoy working with details and coordinating a variety of activities are the key to success. Performers are often people who are very comfortable with developing and implementing the ideas of others.

Figure 6.6 Types of skills needed for each project phase

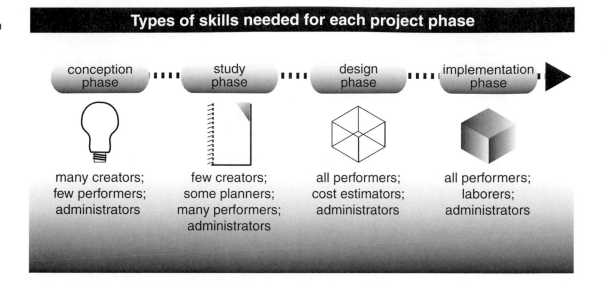

They are usually people who enjoy the detail work. They are rewarded by observing the fruits of their efforts in the form of structures, equipment, software, or procedures.

Laborers: For the Implementation Phase, those persons who enjoy working with their hands and minds, while watching a project convert from paper to reality, are vital. Laborers are individuals who enjoy working primarily with their hands on physical activities.

Administrators: Monitoring and control of the overall project requires individuals who can visualize and remember the final goal. Administrators are often people who are excellent at judging performance and guiding the planners and performers. They are usually skilled at gathering records, evaluating results, and recognizing problems, thus sensing when corrective action is necessary. Working with the manager, they must carefully monitor the forward movement of a project from conception through implementation. They must simultaneously be able to compare the actual work being accomplished with the contracted tasks, schedules, and budgets, as discussed in section 7.5.

Which of these groups of persons is most important to a manager? At first, the creators are very important. In later phases, the planners and performers grow in importance. Once the Conception Phase is well underway, the administrators become very important. When there is work to be accomplished, the laborers are indispensable. Actually, this question is similar to asking, Which is more important—the heart, the brain, or the lungs? The answer is that all are needed and that they must work together as a team.

In some cases, managers have to be personnel developers to persuade individuals to accept different roles for a particular project. This way employees act to fulfill the needs of the team and, at the same time, they have the opportunity to develop new skills and become more flexible and more valuable to the company.

Where several projects are in progress at the same time, there should be a labor pool of employees—the team—that is maintained by a company to avoid continuous hiring and layoffs. A manager must examine the labor pool for the best combination of individuals to perform the tasks in progress or the tasks about to be initiated on a project. Sometimes it is necessary to convince employees of the importance of a team effort so they agree to work on tasks to promote the needs of a project, although the tasks may not match their immediate interest and skill levels. A manager who also enjoys being a personnel developer is often best at this effort. Acquiring new skills encourages employees to become more flexible, more valuable, and more enthusiastic.

Managers may need to obtain the services of vendors to assist when the team members do not have the necessary skills or qualifications to complete the project. Such vendors may be involved in the Design Phase and perhaps also in the Implementation Phase.

The vendors who are chosen to implement the project may not necessarily be the lowest bidders, although in many cases the lowest bidder is chosen.

1. There are times when the lowest bidder may not be the highest in quality and dependability.

2. The lowest bidders may have earned their way onto a bidder's list because they had established themselves as dependable and quality suppliers.

One or more members of a project team may be assigned to investigate potential vendors for their qualifications and recommend which should be used.

How and When Should People Be Evaluated? Personnel are entitled to know how they are performing. Therefore, managers must evaluate their personnel at regular intervals, via performance reviews, so the employees have an opportunity to improve their performance, attitude, and self-image. In a similar way, managers are judged by their superiors; presidents are judged by the company stockholders.

Managers must evaluate their personnel at regular intervals, via performance reviews, so the employees have an opportunity to improve their performance, attitude, and self-image.

There are two types of performance reviews related to projects and programs: **contract performance reviews** and **personnel performance reviews**. The terms used vary from one company to another.

1. A contract performance review involves those who are working on the contract and evaluates how satisfactorily they are fulfilling the terms of the contract and how well they are satisfying the customer. This type of review may be written by an individual who is responsible for the entire contract. Often such a review may result in a memorandum being placed in the personnel file of the involved management individuals. Such a memorandum may be one of congratulations for work well performed; it may be a reprimand for neglecting an important part of the contract or for offending a client.

2. A personnel performance review involves each individual working for a company and that individual's supervisor. It is usually a formal document with a number of topics. The review may involve more than one contract and more than one supervisor. A separate review may be prepared by each of the involved supervisors, or these supervisors may condense their opinions into one document. Each person is judged on the basis of skills and on work as a team member. Personnel performance reviews normally occur regularly at three-, six-, or twelve-month intervals, and also when a significant change occurs in the individual's performance or status.

Pay increases for an individual are usually decided on the basis of performance as reported in both formal and informal reviews. They are recommended by the individual's immediate supervisor and usually require the approval of the personnel department.

People are usually offered promotions when an opening occurs that requires greater responsibility or use of skills on the part of the individual. A promotion may be accompanied by a pay increase or that increase may be withheld until the individual demonstrates the ability to perform in the new position.

One of the major drawbacks of using the matrix style of organization for a project is that the project manager does not evaluate the employees who are brought in from the functional areas. The manager of the functional area (i.e., marketing, engineering) generally evaluates the staff for raises and merit recognition. The employee works for one person, but is evaluated by another. This situation often results in unhappy employees who may look elsewhere for a job.

For projects such as the activity center, pay increases and promotions are not applicable. Students are learning to work together while they gain new skills—both technical and group-related.

SECTION 6.5 PROJECT REPORTING

Well-trained managers are those who have learned to sense when sufficient factors have changed to require project redirection. Information in reports allows managers to know when redirection may be necessary.

Managers need reports so they can choose effective directions for projects. They need accurate and timely information. Managers must realize that, as work on a project progresses, information may change. Well-trained managers are those who have learned to sense when sufficient factors have changed to require project redirection. Reports are often prepared and presented in the form of spreadsheets.

A spreadsheet is the presentation of an organized arrangement of data in matrix (tabular) form. Spreadsheets contain elements organized into rows (horizontal) and columns (vertical) of information. The individual elements or entries are known as cells. It is easy for people to revise the data

Spreadsheet with actual expenditures for six months

month category:	jan	feb	mar	apr	may	jun	totals by category
rent	70	70	70	70	70	70	420
utilities	26	28	29	24	18	15	140
auto	30	32	27	36	41	39	205
food	274	241	269	257	272	245	1558
totals by month	400	371	395	387	401	369	2323

when the spreadsheet is prepared on a computer, and computer programs can quickly manipulate and perform a variety of mathematical operations on those data (Shim and Siegel, 1984).

Expenses may be presented in tabular form. As an example, when a spreadsheet is constructed using a computer program, the rows can be added automatically across (horizontally). Also, the columns can be added automatically down (vertically). Thus, as an entry is changed in one or more data cells, new totals are available quickly and automatically. The updating of the spreadsheet cells must be accomplished on a regular and timely basis if spreadsheets are to be useful for project monitoring. Examine the spreadsheet in figure 6.7; it contains six months of information about actual expenditures.

On the example spreadsheet, the computer software computes the Totals by Month column information and Totals by Category row information. Thus, any changes entered into the data-entry cells cause all the Totals by Month and Totals by Category cells to be corrected to reflect the newly entered data.

This matrix of four rows and six columns contains 4×6, or 24 cells. One can quickly determine that $241 was spent on food during February and $29 was spent on utilities during March. Also, one can quickly determine that the total expenses for June were $369 and that the total cost of utilities for the six months was $140.

Computer spreadsheets are important tools for project managers. Spreadsheets summarize the financial status of a project. Computers allow spreadsheets to be updated quickly and easily. Results are easy to use and can be transferred instantaneously to all those involved in a project.

During the performance of a project, as actual information becomes available, this new information can be entered into the spreadsheet program, replacing the estimated information. As each item of information is entered, the computer software recomputes the spreadsheet and presents new column and row totals on the screen, providing an immediate and up-to-date review of the project status. This information may be printed by the computer equipment as hard copy and distributed to those who are concerned with the financial status of the project. If preferred, this information may be offered via such media as files transferred onto a local area network (LAN), e-mail, or Internet for appropriate personnel to display on their computer screens.

Properly organized information is easy to use, understand, and change. It is easy to examine entries and compare them with previously generated data. In the planning stages of a project, a tabulation of the estimated income and expenses for the project is prepared. This tabulation is known as a balance sheet. A balance sheet that consists of estimates can be automatically updated with actual information on income and expenses, as is discussed below.

Modern computers contain large amounts of memory, and computer programs are flexible. Most programs can be modified by technical personnel to fit the needs of a given company and project. For the activity center project, it was necessary for Manuel to devise a specialized spreadsheet program (Oberlender, 1993) in which data could be quickly entered and displayed, and the cost differentials immediately computed and displayed. Further, it was necessary for Manuel to obtain accurate data weekly from both the Northeastern University Treasurer's Office and from Accounting personnel. Normally, such data are supplied at the end of the month. For the activity center project, that would have been too late for the necessary cost-control requirements.

The problem with updating computer spreadsheets is typically obtaining accurate and up-to-date information to enter into the computer program.

The most frustrating problem for any project group is the need for up-to-date information to be entered into the spreadsheet. The problem is seldom one of computer time and availability. The problem is obtaining data and having the data approved before the information is entered on the spreadsheet. The "garbage in, garbage out" principle suggests that it is better to wait for verified, final information rather than to permit someone to enter interim information that might be incomplete and misleading.

Figure 6.8 A simplified "estimates versus actuals" spreadsheet

A simplified "estimates versus actuals" spreadsheet

		\ multicolumn month category:						totals by category:
		jan	feb	mar	apr	may	jun	
estimates:	rent	70	70	70	70	70	70	420
actuals:	rent	70	70	70	70	70	70	420
difference:	rent							0
estimates:	utilities	25	30	30	25	25	20	155
actuals:	utilities	26	28	29	24	18	15	140
difference:	utilities							−15
estimates:	auto	30	35	30	35	35	35	200
actuals:	auto	30	32	27	36	41	39	205
difference:	auto							+5
estimates:	food	275	240	270	260	270	245	1560
actuals:	food	274	241	269	257	272	245	1558
difference:	food							−2

						totals by month:
estimates	400	375	400	390	400	370	2335
actuals	400	371	395	387	401	369	2323
differences							−12

As noted above, spreadsheets can also be designed to include both the estimated expenditures from the earlier (Conception, Study, and Design) phases, and the actual expenditures as they occur during the Implementation Phase. Figure 6.8 is an example spreadsheet where **estimates** are compared to **actuals.** (Note that estimated values often end in either a 5 or a 0.)

Managers monitoring "estimates versus actuals" can quickly determine if their budgets are within the limits acceptable to them. If there is concern that expenses are exceeding estimates, then the "estimates versus actuals" information can be monitored weekly instead of monthly.

Another solution for immediate expenditure information is to enter data that can be provided by those individuals who are incurring project expenses. These people may be the ones ordering the materials, receiving the materials, receiving the actual bills, or paying these bills. The entries, including labor expenditure entries, should be coordinated through one person and, if possible, coded to note the information source. Why? The answer is that expected costs may change once the vendor's bill arrives. No matter when and how data are entered, every effort must be made to ensure that the data are timely and, at least for the moment, provide enough information to allow for a general control of expenditures.

SECTION 6.6   GATHERING AND ORGANIZING DATA

In this section, application of spreadsheets to monitor and control a variety of data is discussed. Document formats may be standard or unique—that is, prepared especially for the project. The use of electronic databases is covered. Finally, an overview of both corporate and government reporting requirements is given. Additional details on these topics are available in chapters 7 and 8, and appendix F.

The use of a computer provides a fast and reliable method of processing different types of business documents, such as these:

* Personnel records
* Vendor data

- Purchase orders
- Invoices

There are also technical records, such as these:

- Engineering change orders (ECOs) and notices (ECNs)
- Equipment test and evaluation reports
- Development and evaluation procedures

Such documents may contain large amounts of written material—text. However, these documents also contain numerical data that might be extracted for other uses via spreadsheets.

Spreadsheets are valuable documents because a trained user can quickly view essential information. Therefore, many types of documents can be presented in spreadsheet format, such as these:

- Time sheets
- Payroll reports
- Profit-and-loss statements

Predetermined sets of software instructions can be written so that computer programs automatically insert the data entries from one type of spreadsheet into other types of spreadsheets.

Standard document formats may be purchased or a special document format may be developed. See appendix F. A standard document may be modified to suit the needs of a project. As an example, Manuel modified several spreadsheet programs for the activity center project.

Business operations require the preparation, review, and retention of various types of information that relate to individuals, company projects, and the company. Much of this information is available from existing databases, which are collections of related information.

Business operations require the preparation, review, and retention of various types of information that relate to individuals, company projects, and the company. Before computers became cost effective and user friendly, this information was prepared manually (by hand), forwarded to other parts of the company, and stored in several places. Revisions to such information were time-consuming and expensive. Often, changes in information were either not forwarded to the correct place or were not recorded. Reports were prepared that overlapped and often contained conflicting information because of the inaccuracy and age of the data.

Large quantities of both numeric data and text can now be stored and quickly accessed. Advances in both hardware and software capabilities have led to the design of an approach to information storage and retrieval known as a **database.** A database is a collection of related information that can be accessed by a variety of users and that can be updated—modified with later information—as required.

The cost of data storage is decreasing significantly every year so that the size and flexibility of databases continue to grow. A database can be accessed and revised—shared—by authorized individuals. Such a database is known as a **common database.** Reports can be prepared quickly via this common database.

There is information that government agencies insist must be established and retained, such as the following:

- Company income
- Payroll data
- Rent (occupancy) information
- Materials expenditures
- Travel data

These records are provided to government agencies on request or on a routine basis. The agencies may then verify the information by comparing it with the data supplied by others. As an example, the Internal Revenue Service (IRS) compares the company-supplied payroll data with an employee's personal income tax report data.

Figure 6.9 Typical
common database
content

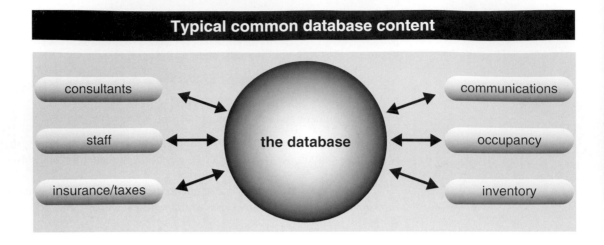

Also, there is information that the company may wish to retain for its own needs, such as the following:

- Evaluating the work that was accomplished
- Preparing cost and schedule estimates for later projects

Using this information, company executives are able to prepare more precise cost estimates and schedules based on the company's past experiences. These documents increase the company's chances of being awarded other contracts while remaining profitable. On the other hand, a bad experience on a certain kind of project may cause the executives to decide not to bid on a project that has similar characteristics.

The common database and related documents for a project might include information regarding the following (see figure 6.9):

- Consultants: individuals hired for brief periods of time for their specialized skills
- Staff: employees, who represent the continuing skills of the company
- Insurance/taxes: liability, workers' compensation, sales, and unemployment taxes
- Communications: mailings, telephone, and copying expenses
- Occupancy: rent, utilities, and maintenance
- Inventory: materials and supplies used while operating a business.

Other information such as the following may also be of value:

- Potential vendors: names of companies that offer skills that the team does not possess
- Office supply sources: names of companies that offer needed supplies at reasonable prices

It is customary to allow modification of the database to be performed by one person or by a very restricted group of persons. The reasoning is that these data should be changed only by someone who is able to verify that the entries are correct and appropriate. It is also customary to control access to the database so it can be used only by persons authorized by the team leader. Thus, it should be difficult for competitors to gain access to information.

A common database would have information useful for all departments in the corporation, such as personnel data or financial records. However, typically not all data can be accessed by all departments.

Leaders of a corporation must be able to extract data from a common database. They want the common database organized to provide appropriate data for the following groups in the corporation:

- Project-control personnel, to ensure that all contracts are profitable
- Accounting personnel, to monitor the project and corporate financial status
- Personnel in the Personnel Department, to maintain employee records

Figure 6.10 Typical structure of a corporate database

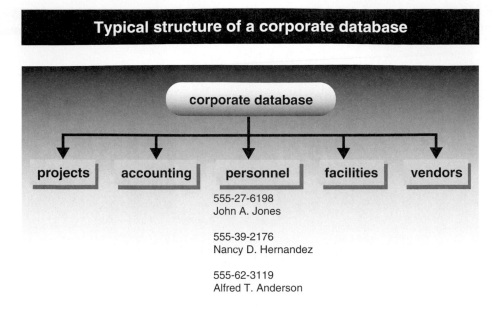

- Facilities personnel, to monitor the cost and usage of the buildings and grounds
- Purchasing personnel, to monitor vendors so they can evaluate their availability, cost, and performance

Records must be designed to provide corporate executives with data in a form that is of value to them.

Figure 6.10 examines how a corporate database could be constructed using the Personnel Department as an example. Note that each of the personnel records for the three people shown in figure 6.10 is identified by a unique number, in this case, their social security numbers. These records are protected by an **access code** so that only authorized personnel can extract information regarding employees. If the corporate officers do not wish their own personal records to be accessed by the authorized personnel, then their records can be further coded to allow access only by a specified corporate officer such as the head of the Personnel or Human Resources Department. Additional detail is given in appendix F.

Assume that Alfred T. Anderson is a salaried employee at a construction company and that he is working part of the time assisting Marketing in searching for new business and part of the time working on the activity center project. The supervisor responsible for the Marketing activity will approve Alfred's charges to his efforts for Marketing. The supervisor of the activity center project will examine Alfred's charges to that project. These two supervisors, or people working for them, must have access to a portion of the database for Alfred T. Anderson.

Employees usually submit work-related data weekly, either using a completed time sheet or time card. These hard-copy time data must be approved by someone—ordinarily by their immediate supervisor. Once approved, these data are entered into the common database.

Prior to the beginning of a project effort, numbers are assigned to each task in that project. These numbers are known as **charge numbers.** As individuals work on a specific task, they record the number of hours worked, along with its assigned charge number, on their weekly time card.

Alfred is issued one charge number for his work on the activity center project and another charge number for his work in Marketing in searching for new business. Because he is working on two unrelated projects, two different supervisors must approve his time charges before they are entered into the common database for automatic data processing.

In addition to the Marketing and activity center project groups, Accounting and Personnel both need access to Alfred's time charges. Accounting prepares his paycheck and Personnel must determine when he has used his allotted vacation or sick leave.

Although Accounting personnel must have access to the payroll portion of Alfred's record, they have no need to know about his personal nonpayroll data. Restricted access codes can be used to prevent unauthorized persons from gaining access to other files and items within

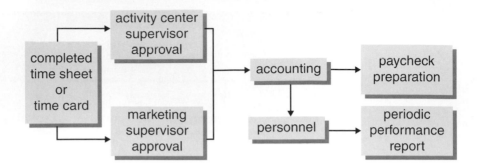

Alfred's record. The Personnel Department normally has access to all of Alfred's record so it can maintain a continuous record of all aspects of his employment history and his contributions to the company. The time-charge information flow is shown in figure 6.11.

Additional details on various records and their interaction are given both in chapter 8 and in appendix F.

SECTION 6.7   PROJECT AND PROGRAM CONTROL

No matter what the status of the effort, or what the latest problems and changes being considered might be, it is the continuing responsibility of the manager to realize that any moment in time is but a small part of the overall work. The manager must continually consider how the present problems and plans affect the entire effort. The manager must not become so lost in the day-to-day work that the primary goals and objectives of the project are forgotten: If you are in a forest, you may not be able to see all the trees.

The manager must continually consider how the present problems and plans affect the entire effort. The manager must not become so lost in the day-to-day work that the primary goals and objectives of the project are forgotten.

On small projects, the manager is directly involved with controlling day-to-day activities and continually monitoring a project's performance.

On larger projects, a project office is established to oversee the proper completion of tasks, adherence to schedules, and performance within budgets. Some person, or group of persons, must be assigned to continually monitor and control project performance. For large projects and all programs, this group of people often has the title of program office. These people perform the following tasks:

• Gather data on task, schedule, and budget, and monitor the total effort

• Analyze and assess progress as compared with the projected schedule and budget

• Recommend the revision of plans where progress is unsatisfactory

• Identify special accomplishments that should be recognized publicly

The program office may also consist of a larger group of people that includes specialists who are responsible for judging the technical work of contributors. Typically, the program office has monitoring responsibility only, without the authority to direct or control the daily operations of the people performing the work. That authority is usually delegated to the line organizations performing the assigned work.

Note that a project office usually has more authority and control over small projects than does a program office. The extent of control given to either the project office or program office depends on the amount of authority and responsibility delegated by corporate management. It is the responsibility of management personnel (corporate, project office, or program office) to monitor and/or control the work that a group of people is to accomplish. It is the responsibility of workers to accomplish this work.

Figure 6.12 Monitoring and Control

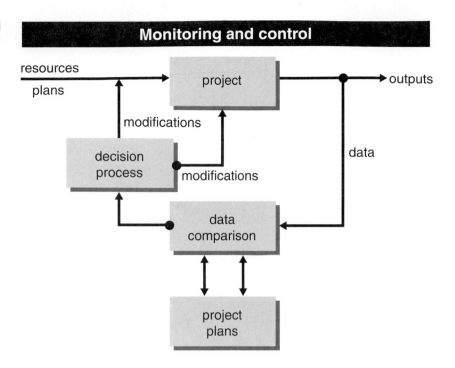

Monitoring and control

resources

plans → project → outputs

modifications

decision process

modifications

data

data comparison

project plans

Control of a project or program is accomplished when the project managers or program managers issue instructions that ensure better understanding, coordination, and follow-through by the workers.

To ensure control, the project or program manager issues directives—orders. These orders are usually in the form of new and revised task descriptions, including staff, schedule, budget assignments, and charge numbers. Control of a project or program is accomplished when the project managers or program managers issue instructions that ensure better understanding, coordination, and follow-through by the workers.

The monitoring and control process can be viewed as an elementary control system. As illustrated in figure 6.12, the control process starts with collecting data from the project. The progress being made on the project, as reflected by the data, is then compared to a standard or a project plan. The results of these comparisons are evaluated by the project manager or assigned senior team members. A decision is made either to modify the work plan or leave it alone. If the decision is to modify the work plan, then the changes are recorded on the project plan so that, at the next review, the correct comparisons can be made.

The project initiation meeting is the name often given to the first meeting of those who will be involved with a project. This is the time to acquaint all of those persons on the project with where they fit into the overall effort. Written task descriptions (see section 7.2 and appendix B), schedules, budgets, and charge-number lists should be delivered to the invited participants prior to the meeting. The project plan and schedule must be presented so that all participants know where they will start and what is expected of them. Questions are often asked; not all of these questions will have immediate answers, however.

Managers typically review projects in portions, such as tasks and subtasks. Projects may be reviewed more frequently when the manager feels that such reviews are necessary.

Managers must review the total project or program periodically. Reviews for a large project may include only team leaders or supervisory personnel. These people then report the results of the meeting to their team members.

A project is typically reviewed in portions, such as tasks and subtasks. It may be reviewed more frequently when the manager feels that such reviews are necessary. Think of the changes in a project as small as the activity center project. During its performance, the amount of available money changed. Also, the sequence of the work to be accomplished had to be changed to match the schedules of contractors, suppliers, and student workers.

Reviews should begin during the Conception Phase, at that point when progress can be measured or when decisions must occur regarding changes in direction. Reviews should always occur at the end of a phase so that all persons involved in the next phase may become aware of what has been decided, what was accomplished, and what questions remain to be explored. Persons presenting information at these reviews must be well prepared to answer questions from the meeting attendees. These persons should be flexible in presenting their planned presentations;

Figure 6.13
Sequence for project
reviews

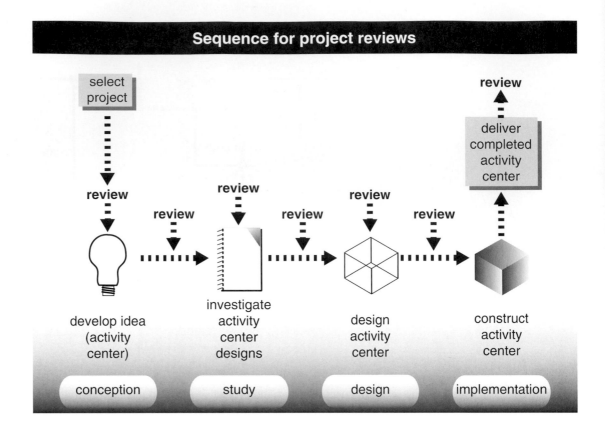

often a question may lead to an unexpected path of discussion. The project personnel participating in the review should speak in a neutral manner so that they will not appear defensive of
their views but are willing to accept constructive criticism.

A typical project sequence, showing when reviews might occur for the activity center, is
given in figure 6.13.

It is important to schedule a review after some milestone in a project has been achieved.
Many companies schedule their project reviews according to the number of weeks after program start or according to some fixed time period such as the end of the month. Performing reviews at fixed time intervals is convenient for scheduling the review process, but it often leads
to redundant reports and duplication of effort. Focus on the project deliverables. Schedule project reviews at points in the project where adjustment in resources may be required. This approach is much more effective than specifying a weekly, monthly, or quarterly review.

*Managers have financial con-
trol of projects. They need to*

- *Ensure that funds are available when needed*
- *Ensure that work is completed within the estab-
lished budget*

The responsibility for the project or program finances is normally assigned to its manager.
These responsibilities include the following:

- Ensure that funds are available when needed.
- Plan in advance so a budget surplus exists should extra funds be required later.
- Establish a cost center where cost control can be performed.
- Delegate financial responsibility for daily or weekly control to assigned personnel.
- Ensure that the work is completed on time and within the established budget.

Managers are judged by their ability to complete the effort on time, within budget, to the preagreed
specifications, and to the satisfaction of the client. At the completion of a project, the manager must
generate enough earnings or profit so that the company can survive to bid on other contracts.

When a project is nearing completion, there is a large amount of documentation, known
as paperwork, that needs to be prepared, distributed, and filed for future reference. The project
is not complete until the paperwork is finished.

Typical documentation includes the following:

- Engineering and manufacturing reports
- Test reports
- Instruction manuals
- Financial statements
- Cost/volume/profit analyses
- Overhead and borrowing costs distribution
- Return-on-investment report

These and other contracted reports or company-required reports must be prepared and reviewed by the manager's staff. Therefore, funds must be reserved to pay for the generation and review of this documentation. See chapters 7 and 8.

CHAPTER OBJECTIVES SUMMARY

Now that you have finished this chapter. you should be able to:

1. Explain the four key responsibilities of managers.
2. Discuss how the role of manager changes over the phases of a project.
3. Describe a typical organizational structure, which includes staff, line, and support personnel. Discuss the role of each group in the organization in project planning, performance, and implementation.
4. Describe the two main styles of management and the advantages and disadvantages of each.
5. Discuss ways managers consider skills and interests of individuals in choosing people to assume certain roles in a project.
6. Explain why performance reviews are important. Describe criteria on which people can be evaluated.
7. Define spreadsheets and explain how they are useful to managers.
8. Define common databases and explain how they are used by businesses.
9. Outline the responsibilities of project and program managers during the performance of a project and program. Describe how they can review and control the project and program.

EXERCISES

6.1   Talk to students who have performed technical work during the summer or while on a cooperative education assignment to determine why management is necessary. Determine how the organization is divided into staff and line groups, and how consultants are used. Ask them about and then list the positives and negatives regarding management at the locations where they were working. Also have them analyze how an improvement in management could have saved time and money.

6.2   Obtain or sketch an organization chart from the students interviewed in exercise 6.1 above. Identify the staff and line groups.

6.3   List any specialty consultants involved in exercise 6.1 above. Identify whether they were from within the organization or were hired as outside consultants. Identify their specialties.

6.4   From exercise 6.1, describe whether the management practiced top-down management, bottom-up management, or a combination of both. Note your opinion of the advantages and disadvantages of the management style applied as related to the work being managed.

6.5   Develop a performance review document that could be applied to evaluate the work of an employee. Also prepare three additional parts: one for a professional employee, one for a laborer, and another for a consultant.

6.6   Examine pay raises from the point of view of both the employee and management. For this exercise, ignore the effects of inflation. For a person who works for the same company for more than twenty years, calculate the salary of that employee at the end of that time, assuming annual 5 percent pay raises.

6.7   Discuss how you, as a manager, would react to an employee who had not taken advanced courses and yet expected a promotion or a significant pay increase. Also discuss how much more capability you would expect from an employee who had received 5 percent pay raises every year for twenty years.

6.8   Examine the skills of an entering technical employee. Discuss when you would consider that employee to have earned a promotion. For exercise 6.1 above, obtain a list of technical positions, their titles, and (if possible) a job description.

6.9   Examine at least two spreadsheet computer programs. Describe the advantages and disadvantages of each, particularly regarding how each spreadsheet is updated and how that updating affects the common database and presentation of estimates versus actuals.

6.10   Discuss how you would react to a project office or program office regarding how closely their members would want to control your daily work. Then discuss how you as a member of one of those offices would react if the daily work was not being accomplished on time and within budget.

6.11   Assume you are assigned the duties of a project manager. Develop a two-part plan to

1. improve your managerial skills
2. maintain your technical skills

Also realize that your assignment may require an extensive amount of your working day.

6.12   Assume you are assigned the duties of a project manager. The program manager in charge of your project directs you to focus on efficient and timely project completion. At the same time, your department manager directs you to apply your team members in such a manner that their technical skills increase. Develop a plan that will satisfy both of your managers.

6.13   Discuss your options for reassigning project personnel on completion of your project.

6.14   Write a paper describing what you have learned from your project work. Indicate how your involvement in this work has affected your choice of a professional career, if applicable. Revise your resume to reflect your newly acquired skills and experience and, if necessary, your career-change direction.

chapter 7

Project Management

The Project Plan

Specifications and Reports

*Modeling and
System Design*

Project Management

The Project Plan

Specifications and Reports

Modeling and System Design

1. Can you explain the importance of project planning?

2. Are you able to describe and develop a Work Breakdown Structure?

3. Have you ever created a Gantt chart for a project?

4. Can you establish who is responsible for project tasks and, with their assistance, create a responsibility matrix?

5. Can you build a network diagram for a project?

6. Are you able to analyze a network diagram and establish an acceptable schedule for the work?

7. Can you list the various kinds of costs that must be monitored in a budget?

8. Do you understand the schedule adjustment process and how to use software in making the adjustments?

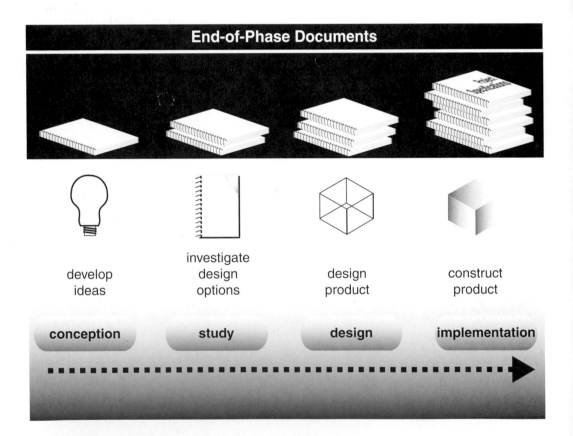

End-of-Phase Documents

develop ideas

investigate design options

design product

construct product

conception   **study**   **design**   **implementation**

What is conceived well is expressed well.

Nicolas Boileau, 1674

Planning, scheduling, and controlling is often referred to as the project management cycle. Each of these phases of the management cycle are closely linked together. One objective of this chapter is to illustrate approaches for developing schedules and budgets, and for monitoring and controlling a project.

One of the first of many challenges faced by a project manager is the development of a project plan. A project plan is not a single document. It consists of a Work Breakdown Structure, a responsibility chart, a schedule in the form of a bar chart and a network diagram, and supporting documentation describing costs and budgets.

The manager and the members of the project team prepare a project plan as a road map to provide them with direction and guidance from project kick-off to the day of delivery. The length and complexity of the project plan depend entirely on the type of project and the level of investment being made in it. For example, a real estate developer might only prepare a Work Breakdown Structure and a schedule of subcontractor due dates when building a single house. However, when the same contractor is building a 300-unit condominium complex, the work is broken down to a level of detail that will assist in developing detailed schedules and responsibility matrices.

After you complete the study of this chapter, you should know how to do the following:

- Participate in the preparation of written specifications.
  Subdivide a project into a hierarchy of tasks and work packages.
  Write clear task descriptions.

- Prepare a Work Breakdown Structure and the associated documentation.
- Create a responsibility matrix.
- Develop a Gantt chart for timing the project.
- Draw a network diagram for analyzing the project.
- Apply the Critical Path Method of analysis to develop a schedule for the project.
- Adjust a project schedule to match customer needs.
- Understand the budgeting process.
- Realize the complexities associated with the interaction of tasks, schedules, and budgets.
- Use a software package to organize data.

SECTION 7.1   THE WORK BREAKDOWN STRUCTURE

The Work Breakdown Structure (WBS) is often the first document developed as the project manager and project team formulate a project management plan and schedule. The schedule defines the sequence and duration of tasks that must be done to complete the project on time and within the cost guidelines established. Developing a Work Breakdown Structure is usually an iterative process. At each iteration new personnel get involved and more detail is developed about the activities, actions, processes, and operations that need to be performed to satisfy the customer's requirements.

Sometimes a company performs almost identical types of projects for every customer. Therefore, the project manager has an excellent grasp of the tasks required and the personnel who can perform these tasks. In these cases the Work Breakdown Structure may be nothing more than a standard form completed by the project manager. However, such a situation is very rare. Most projects are one-time occurrences, each with a unique set of characteristics that ultimately leads to a well-defined, but very individualized, set of results.

The purpose of the Work Breakdown Structure is to demonstrate clearly to all parties involved how each task is related to the whole project in terms of budgets, schedules, performance, and responsibility for the physical assets belonging to the project. Some project managers include enough detail so that the Work Breakdown Structure can be used as an instrument for scheduling, personnel assignments, resource allocations, monitoring project progress, and controlling how and when tasks are accomplished.

The objective of the Work Breakdown Structure is to partition the project into distinct work packages, each of which contains detailed work elements. The Work Breakdown Structure starts as a project task list, usually an indented list or a node tree, that shows the relationship of the tasks to each other. The indented list in figure 7.1 is a hierarchical list of tasks that must be performed to complete the project. The indented position of a task in the list indicates its position in the hierarchy of tasks. The more a task's position is indented, the further down it is in the hierarchy. The node tree shown in figure 7.2 is a hierarchical listing of the functional areas of the organization required to complete a project.

The Work Breakdown Structure can be developed by either the entire project team or a subset of the team. The composition of the team depends on the complexity of the project and the type of knowledge required. However, as the document is drafted, the following features must be included in the final package:

1. Each project task that forms an element of the Work Breakdown Structure should be big enough to be identified and assigned to an individual member of the project team.

2. Each task should be defined clearly enough that it can be viewed as a package of work to be accomplished by the project team.

Figure 7.1 Indented Work Breakdown Structure (WBS): partial task list

Indented Work Breakdown Structure (WBS): partial task list

```
1.0    purchase land (task)
       1.1    search for land (subtask)
       1.2    assess quality of land
       1.3    finalize purchase of land
2.0    prepare site (task)
       2.1    mark and stake lot lines
       2.2    cut and remove brush
       2.3    level lot
              2.3.1   bring in dirt
              2.3.2   grade lot
3.0    excavate site (task)
       3.1    stake foundation
       3.2    dig foundation hole
       3.3    stake position of footings
4.0    erect foundation (task)
       4.1    prepare footing frames
       4.2    pour footings
       4.3    install foundation forms
       4.4    pour concrete
       4.5    waterproof exterior walls
       4.6    allow foundation to dry
       4.7    backfill completed foundation
```

Figure 7.2 Node tree or tree diagram

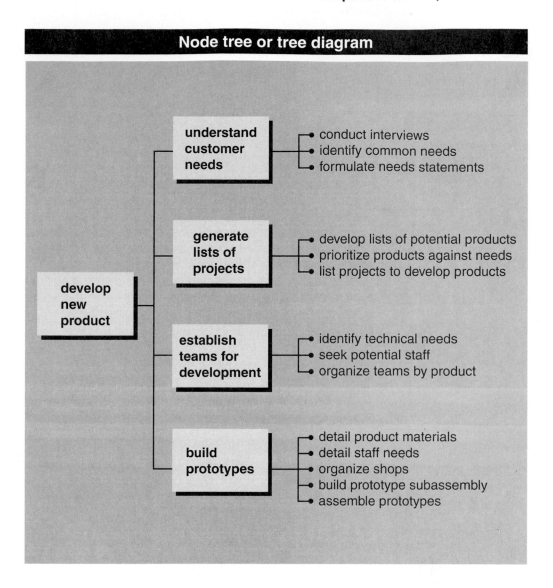

Node tree or tree diagram

3. Each task should be described in such a way that it is possible to define each element of the task in terms of the money, labor, and resource materials required.

As the team develops the Work Breakdown Structure, it is vital for them to remember that this package of documents will form the foundation for all future work. After work has actually begun, it is frustrating to discover that an essential task has not been considered. Replanning a project and developing engineering change orders are both difficult and expensive. A careful and diligent effort to draft a comprehensive plan initially will reap major dividends. The project manager should never underestimate the value of developing a thorough understanding of the project right from the beginning.

As the Work Breakdown Structure evolves, a questioning attitude should prevail. Those involved should carefully consider not only *what* needs to be done, but also *why* it is necessary. "What" and "why" questions are critical elements in forcing the team to focus on the most important activities associated with project completion. This focus prevents distraction by those elements that may be unnecessary or even trivial to success. "Where" and "when" questions compel the team to identify those places where the work will take place and the optimum times during which it will be done. Defining who will do the work and asking how they will do it refines and drives an understanding of the tasks to a more precise level of detail.

A general set of steps to follow when preparing a Work Breakdown Structure is as follows:

1. Have the people who will be responsible for the tasks prepare a list of work units. Keep asking the questions: What needs to be done, and Why?

2. Organize the tasks in the order in which they must be completed and continue to break the tasks down into more precise levels of detail. Work hard at establishing the beginning and the ending points for each individual task. Remember that the task description must facilitate the following vital elements: scheduling, budgeting, monitoring, and controlling the project.

3. Organize the work packages into a hierarchy of tasks. This can generally be done with the aid of a node tree diagram. Complete the following steps for each work element:

    a. Write a work description that includes data inputs, materials requirements, contract requirements, and any unique features associated with completing the work.

    b. List the suppliers, vendors, and personnel associated with the tasks. To the extent possible, specifically identify individuals and vendors.

    c. Identify the tools required to perform the work. These include types of equipment, software packages, or unique skill sets requisite to personnel.

    d. Specify the output or the products of the task (activity) such as reports, design specifications, hardware, or other specific items.

    e. Develop a list of personnel who will be doing the work associated with each task.

4. Create an indented list of tasks including the task description, the inputs to the tasks, the outputs from the tasks, the mechanisms or tools required, and personnel responsible for the work. Figure 7.3 illustrates such a list.

Work Breakdown Structure with details

activity		responsible	input	output	time
1.0	purchase land	team	listings	owned land	
1.1	search for land	Tom			5 wks
1.2	assess land quality	Tom, realtors			2 wks
1.3	finalize purchase of land	lawyers			2 wks
2.0	prepare site	surveyors	site topology	marked plan	
2.1	mark and stake lot lines	surveyors			1 wk
2.2	cut and remove brush	laborers			2 wks
2.3	level lot	excavator			2 wks
2.3.1	bring in fill dirt	excavator			1 wk
2.3.2	grade lot	excavator			2 wks
3.0	excavate site	excavator	equipment	prepared site	
3.1	stake foundation	surveyor			1 wk
3.2	dig foundation hole	excavator			2 wks
3.3	stake position of footings	surveyor			0.5 wk
			contracts		
4.0	erect foundation	concrete contractor		completed foundation	
4.1	prepare footing frames	laborer			0.5 wk
4.2	pour footings	concrete contractor			0.5 wk
4.3	install foundation forms	laborer			0.5 wk
4.4	pour concrete	concrete contractor			0.5 wk
4.5	waterproof exterior walls	contractor			0.5 wk
4.6	allow foundation to dry	none			3 wks
4.7	backfill completed foundation	excavator			1 wk

Figure 7.3 Work Breakdown Structure with details

SECTION 7.2   ESTABLISHING RESPONSIBILITY FOR TASKS

A major goal of project planning is assigning responsibility for the various elements of the project before the project is even started. The project manager who attempts to assign tasks as the project progresses is at serious risk of failure.

Each person who needs to see the work or bears responsibility for completion of the work must be identified in the project management plan. Although the plan may assign the performance of a task to a particular group of individuals, others will generally be involved to some degree. For example, a project to write a proposal for funding from a government agency may initially involve two or three members of a project team. Ultimately, however, the proposal may be checked and edited by the project manager, reviewed by the division director, and evaluated for approval by the vice president of research.

A responsibility matrix is a particularly useful technique for associating a list of individuals who will be involved in the project with the tasks that must be completed. Although there seem to be as many methods of presenting a responsibility matrix as there are project management consultants, all such matrices contain essentially the same information. The basic responsibility matrix is a grid whose left-hand column lists the tasks or work packages defined in the Work Breakdown Structure. Along the top of the grid, project team members, vendors, and others responsible for various tasks are identified by job title or name. The matrix is completed by inserting a code at the intersection of task and person that indicates the person's type and level of responsibility for that task. Table 7.1 illustrates a responsibility matrix for a small project.

A responsibility matrix can be time-consuming and expensive to prepare, but it is time and energy well invested. The matrix gives the project manager a clear indication of who will be responsible for the various tasks of the project. It identifies the levels of management that will be involved and their contribution to the work. If the project manager can obtain agreement from the individuals identified on the responsibility matrix, misunderstandings with respect to task assignments or task ownership can be avoided.

A secondary benefit is also derived from a responsibility matrix, because it highlights the workload assigned to each individual, group, or organization. The project manager can more easily pinpoint those who may compromise the success of the project because they are either overloaded or underutilized.

For a large involved project, the responsibility matrix can become quite complex. Meredith and Mantel (1995) provide illustrations of what they have named *linear responsibility charts*. One reference is a "verbal responsibility chart," which lists activities, actions, the action initiative, and the responsible individuals. In this illustration Meredith and Mantel mention that the original document is thirty pages in length and contains 116 major activities.

Table 7.1   Sample responsibility matrix

WBS Activity Task Listing	Responsibility for Task			
	Project Manager	**Administration**	**Crew Chief**	**Contractor**
1. Purchase land	5	1	5	0
2. Prepare site	2	3	1	0
3. Excavate site	2	4	2	1
4. Erect foundation	2	5	1	1

Key to Responsibilities
  0 = None
  1 = Accomplish
  2 = Supervise
  3 = Must be notified
  4 = Final approval
  5 = Consult for advice

SECTION 7.3   THE PROJECT SCHEDULE

The schedule defines specific points in a project when the project manager must bring in personnel and equipment resources to complete required tasks on time and within budget. Developing a schedule for a project involves deciding when each task and subtask listed in the indented Work Breakdown Structure should begin and how long it should last. The scheduling process translates the list of project tasks into a timetable of events or completion dates.

The scheduling process is similar to the development of a Work Breakdown Structure in that a high-level master schedule to provide a skeleton for the project is constructed and reviewed. High-level tasks that require a more detailed breakdown are selected, and detailed schedules are developed. The timing of these subschedules must fit within the time limits established by the master schedule.

The process of scheduling a project often involves trade-offs.

It is important to realize that the process of scheduling a project often involves making tradeoffs between the cost of personnel and equipment resources and the adjustment of task completion dates to satisfy customer requirements. For example, a project task's duration can be shortened by scheduling overtime work, but the additional labor cost may be too much for the budget to absorb.

Because a project schedule provides a visual representation of the relationships among the different pieces of a project, the method chosen to display these relationships is very important. It should be easy to follow and should not add complexity to an understanding of the project. The Gantt chart or bar chart and the network diagram are two commonly accepted methods for illustrating the relationships among the project elements. These methods result in two different types of graphical representations of the project and are very easy to understand once they are complete. Both of these graphical tools are important and should be used together when building a schedule for a project that goes beyond five or six activities.

The Gantt Chart

The Gantt chart or bar chart is the oldest and most frequently used chart for plotting work activities against a time line. Gantt charts show how long a project will take when tasks that are independent of each other are performed simultaneously.

The most common reason for using a Gantt chart is to monitor the progress of a project. Gantt charts provide an excellent method for comparing planned work activities with the actual progress made on a project. Note how information is organized on the Gantt chart shown in figure 7.4:

- The time in weeks is shown across the top.
- Each project activity is bracketed by inverted triangles ($\triangledown$) that align with the week numbers of the activity start and complete dates.
- Solid lines connecting the inverted triangles show an activity's duration in time.
- Dashed lines improve the readability of the chart.

A Gantt chart similar to the one in figure 7.4 may be used as an overview for related schedules that include more detail. More detailed schedules would include the titles of the individual tasks and subtasks (Callahan, Quackenbush, and Rowings, 1992; Farkas, 1970).

Although it is better to remain with the original overall schedule and completion date, most schedules have to be revised as work on a project progresses. Noninverted (upright) triangles ($\triangle$) on the Gantt chart indicate new planned start and finish dates. If, for example, the designers of the project in figure 7.4 discover that an existing design from a previous project can be modified, they can shorten the "design product" task by one week. However, because of their experience with the previous design (the one being modified), the team members also realize that the construction phase will require an additional week. Therefore, they revise the schedule as shown in figure 7.5, and the triangles move to the left (earlier start or finish), remain where they are (no change), or move to the right (later start or finish).

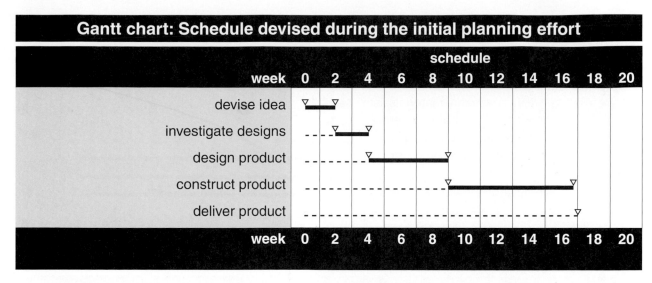

Figure 7.4   Gantt chart: Schedule devised during the initial planning effort

The schedule is modified using the upright triangle notation so that changes from the original schedule are readily apparent. If proposed revisions to the schedule are not accepted by the project manager or client, or if it is not possible to change the proposed schedule, the chart can be readily changed back to what it was before.

Gradually the schedule document will include more detail as the task descriptions are expanded to include information regarding materials and labor skills. For each approved task, rough costs are estimated and verified to the extent possible. An example of a more detailed schedule, a construction schedule, is given in figure 7.6.

Examine the tasks listed on the chart. Note that each task starts with an active verb (for example, *verify, prepare, inspect*) and that the schedule headings are the overall task titles from the Work Breakdown Structure, shortened where necessary. Task numbers may also be included. Several changes have been made to this detailed schedule, and the schedule scale has been enlarged so that calendar dates can be inserted if desired. Items that occur repeatedly, such as "inspect work in progress," are indicated as inverted triangles on a dashed line. The construction schedule now shows a number of tasks to be performed. Some tasks are listed in sequence; the schedule lines of others actually overlap.

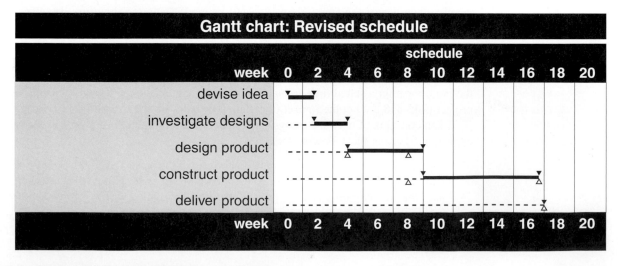

Figure 7.5   Gantt chart: Revised schedule

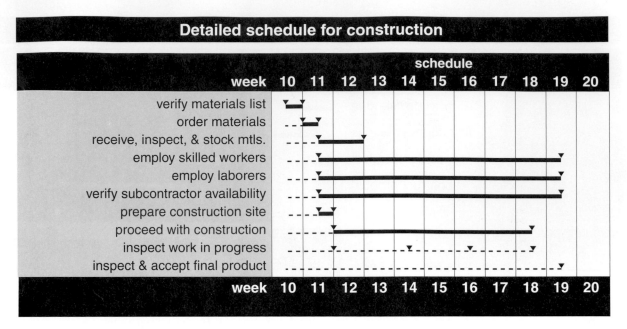

Figure 7.6   Detailed schedule for construction

The principal problem with the Gantt chart is that it does not provide enough information about how various tasks relate to each other. The Gantt Chart can be used by itself to create a master schedule for very small projects. Project managers encounter problems, however, when they try to create a schedule for a more complex project. Some of the following questions may arise:

- What parts of the tasks can be done at the same time?
- Is there any slack time or room for adjustment in the start times for the tasks?
- Are there precedent relationships for the whole task, or do they apply to just parts of the task?

Simply stated, bar charts are awkward to handle. When used in conjunction with a network diagram, however, the bar chart can be very useful.

The Network Diagram

Network diagrams combined with a set of systematic analysis procedures serve to remedy the shortcomings of the Gantt chart. Because network diagrams show timing relationships between various tasks on a project, they form an excellent framework for planning, scheduling, and monitoring the progress of a project. They are also communication tools that allow the project manager to show dependencies among tasks and prevent conflicts in timing. By using an algorithm, a project manager can calculate milestones and determine the probability of completing the project within a stated time frame. The manager can also experiment with starting dates for project tasks and predict the impact of certain start dates on the overall project completion date.

The concepts of network analysis and network representation of a project are more easily explained by using the following terms:

Activity: A task or job that forms a component of the project. Activities consume resources and always have definite start points and endings. This presentation of network diagrams uses the "activity-on-arrow" approach, that is, all activities will be shown as arrows.

Nodes or Events: The beginning and ending points of activities. A node cannot be realized (executed) until all activities entering it are complete. A node is shown as a circle with a label.

Merge Node: A node where more than one activity is completed.

Burst Node: A node where more than one activity is initiated.

Network: A graphical representation of a project plan with the activities shown as arrows that terminate at nodes.

Path: A series of activities joined together to form a distinct route between any two nodes in a network.

Critical Path: The limiting path(s) through the network that defines the duration of the project. Any reduction of the project duration must begin on the critical path.

Precedence Relationship: The order in which activities should be performed. For example, digging a hole for a foundation must precede the pouring of the foundation. There is a precedence relationship between the excavation and the cement-pouring tasks.

Dummy Activities: Fictitious activities that carry a zero time estimate and no cost are used to illustrate precedence requirements in a network diagram.

Time Duration: The total elapsed time of an activity from start to finish. The Critical Path Method (CPM) has one estimate of activity duration. The Program Evaluation and Review Technique (PERT) has three estimates of duration for each activity in the network.

In the activity-on-arrow representation of a project shown in figure 7.7, nodes (circles) are placed at the tip and tail of the activity arrow to identify the beginning and ending events for an activity. In the literature, authors often refer to the arrows as *arcs* and the nodes as *events*. Notice that a dummy activity has been inserted between nodes 3 and 5, showing the precedence requirement between activities B and G.

The following conventions must be observed when developing a network diagram for a project:

1. Each activity is represented by one and only one arrow.
2. All activities begin and end in events (nodes shown as circles on the diagrams).
3. The length of an arrow has no meaning with regard to the duration, cost, or importance of an activity.
4. Arrows originating at an event (node) can only begin after all activities terminating at the event have been completed or, as some authors state, the node has been realized.
5. The first node of the diagram and the last node of the diagram may have only output arrows and input arrows, respectively. All other nodes must have at least one input arrow and one output arrow.
6. No arrow can start and end at the same node.
7. If one activity takes precedence over another, but there are no activities relating them, a dummy activity can be used. The dummy activity will be shown as a dashed line arrow with no duration or cost. (See figure 7.7.)

Figure 7.7 Activity-on-arrow network

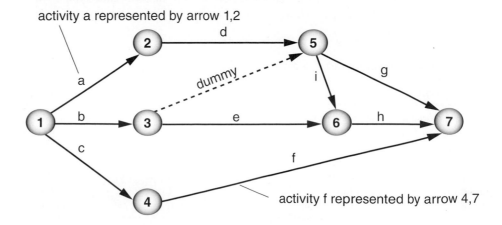

Activity-on-arrow network

Figure 7.8 Network diagram for site preparation project

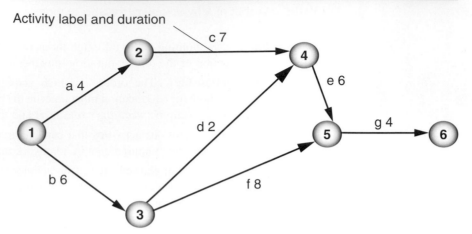

Network diagram for site preparation project

Activity label and duration

Figure 7.8 is an activity-on-arrow representation of the site preparation project described by the Work Breakdown Structure in table 7.2. (The **terminal node** is labeled: 6.) Notice that only high-level functions are modeled in the network diagram. Project managers very often develop a master schedule for the project from the high-level functions. Then they produce sub-schedules for the subactivities that fall within the timing established by the master schedule.

Follow these steps to develop a project network diagram:

1. Make a list of the activities that will drive the master schedule, that is, the activities on the Work Breakdown Structure that form major categories of work.

2. Label these activities (for example, A, B, etc.) so that the diagram is easy to read (see figures 7.7 and 7.8).

3. Determine or estimate the duration of the activity.

4. Note all precedence relationships and check that these relationships are correct.

5. Start with the initial node, label it node 1, and move forward, constructing the project network while continually checking the precedence relationships. For example, developing a network for the project described in table 7.2 will result in the diagram shown in figure 7.8.

The two network diagramming techniques to be discussed in this chapter are the Critical Path Method (CPM) and the Program Evaluation and Review Technique (PERT). CPM was developed by the industrial giant DuPont, Inc., for scheduling maintenance in their chemical plants. PERT was developed by the consulting firm Booz-Allen and Hamilton under contract to the U.S. Navy as part of the Polaris Program.

PERT and CPM both require that the same type of network diagrams be developed for a project before any analysis can be performed. Very often in the literature, the PERT and CPM

Table 7.2 Activities for site preparation project

Activity Description	Activity Label	Duration	Preceding Activities
Measure plot and position stakes for foundation	a	4	none
Identify personnel and equipment needs	b	6	none
Measure perimeter points and bring in fill dirt	c	7	a
Select and assign crew from existing employees	d	2	b
Secure and perform maintenance on equipment	e	8	b
Dig foundation hole and prepare site	f	6	c, d
Grade lot and prepare site for construction	g	4	e, f

techniques will be grouped under the single heading of "Critical Path Methods." The main difference between the two techniques is the method used for calculating task duration times. CPM requires that one fixed time-value estimate be provided for each task. The CPM analysis results in a single time-value estimate of the time to complete the entire project. PERT has been formulated to deal with uncertainties or probabilities of the type found in research-and-development projects. PERT uses three time estimates to calculate the duration of a task. The PERT analysis predicts the completion time of a project.

Before discussing each of the project planning tools in detail, it is important to mention that none of these tools is absolutely mandatory for a project manager to develop a good project management plan. Some companies and government agencies prescribe project management strategies that do not involve the development of responsibility charts or network diagrams. However, the diagramming and analysis tools that are presented in this chapter are used by a very wide range of project managers and companies. In their text titled *Analysis and Control of Production Systems,* Elsayed and Boucher (1985) cite a study that found as many as 80% of 400 construction firms were using the critical path method. In another study, a large percentage of the firms using quantitative methods were using PERT or CPM.

Many useful software packages have been developed to apply the PERT/CPM approaches to a project schedule. See Tapper (1998). Almost daily, software houses announce the availability of inexpensive, user-friendly personal computer (PC) based packages for managing projects. Each year *PC Magazine* reviews the project management software available, much of which is PERT/CPM based.

The Critical Path Method (CPM)

The critical path through a project network is the path that defines the duration of the project. The critical path is the limiting path or the longest path found when examining all potential routes from the initial (first) node to the terminal (last) node. The Critical Path Method (CPM) of project analysis and scheduling provides a stepwise procedure for combining times for activities that must be accomplished and for analyzing precedence relationships. In order to apply CPM, the following definitions must be established.

ES(X): The Earliest Start time for activity X.

tx: Task (activity) duration time

EF(X): The Earliest Finish time = the Earliest Start time plus the task duration time. For activity X with a duration tx, $EF(X) = ES(X) + tx$.

LS(X): The Latest Start time for a task is the latest time that the activity can start without delaying the project.

LF(X): The Latest Finish time for an activity that is started at the Latest Start time. For activity X, $LF(X) = LS(X) + tx$.

EOT: The Earliest Occurrence Time is the earliest time that an event (node) can be realized. When an event has only one preceding activity, the EOT for the event is equal to the earliest finish time for the single preceding activity. When more than one activity enters a node, all activities must be completed before the node is realized. The EOT is the longest time required for all paths leading to the node to be completed.

LOT: The Latest Occurrence Time is the latest time that an event can be realized without delaying the project. The LOT for an event is defined by the activities leaving a node.

To calculate the ES( ) and EF( ) activity times and the EOT event times, we perform what are referred to as **forward-pass calculations.** For these calculations the CPM algorithm assumes that node 1 represents time zero—that is, the beginning point for the project. The ES( ) time for any activity starting at the first node is zero. For the site preparation example illustrated in figure 7.8, ES(a) = ES(b) = 0, 0. The EF( ) for the activities starting at node 1 are calculated by adding the task duration to the earliest start times:

$$EF(a) = ES(a) + ta$$
$$EF(a) = 0 + ta = 0 + 4 = 4$$
$$EF(b) = 0 + tb = 0 + 6 = 6$$

Since node 2 can be realized at time 4, then the EOT for node 2 is 4. The earliest start ES(c) for activity c is 4 and the earliest finish EF(c) for activity c is calculated as $4 + 7 = 11$.

The EOT for node 4 is determined by calculating and comparing the earliest finish times for all activities entering the node. EOT cannot occur until all entering activities have finished. In this case, EOT for node 4 is 11 since that time is the longer of EF(c) and EF(d). This relationship is expressed as:

$$\text{EOT for node 4} = \max\left[EF(c), EF(d)\right]$$

where

$$ES(c) = 4 \text{ and } EF(c) = 4 + 7 = 11$$
$$ES(d) = 6 \text{ and } EF(d) = 6 + 2 = 8$$
$$\text{EOT for node 4} = \max\{11, 8\} = 11$$

$$ES(e) = 11, \text{ resulting in the following:}$$
$$EF(e) = 11 + 6 = 17$$
$$EF(f) = 6 + 8 = 14$$
$$\text{EOT for node 5} = \max\{17, 14\} = 17$$

Therefore

$$ES(g) = 17$$
$$EF(g) = 17 + 4 = 21$$

The final duration for the network is 21 time units with an EOT for node 5 of 17 time units and an EOT for node 6 of 21 time units. Recall that the critical path is the longest path through the network and defines the duration of the project. The critical path for this example consists of activities a, c, e, and g.

In order to calculate the LS( ) and LF( ) activity times and the LOT event times, you must know the duration of the network. The terminal node of the network will define the latest finish time for activities ending at that node. To calculate the times for LS( ), LF( ), and LOT we perform **backward-pass calculations.**

The backward-pass calculations are required to determine the values of the latest start, LS, and the latest finish, LF, times for the project activities. The latest occurrence time, LOT, for the nodes is similarly calculated.

The relationship between the LF and the LS is

$$LF = LS + t$$

Therefore, for our example, the latest finish for activity g and the finish time for the project is 21 time units. Therefore

$$LS(g) = LF(g) - t$$
$$LS(g) = 21 - 4 = 17$$

The LOT for node 5 is equal to 17 since the latest that node 5 could occur and still not delay the project is time 17.

Moving backward through the network:

$$LF(e) \text{ and } LF(f) = 17$$

Table 7.3 CPM calculations for site preparation project activities

Activity	Duration (t)	ES	LS	EF (ES + t)	LF (LS + t)	Slack
a	4	0	0	4	4	0
b	6	0	3	6	9	3
c	7	4	4	11	11	0
d	2	6	9	8	11	3
e	6	11	11	17	17	0
f	8	6	9	14	17	3
g	4	17	17	21	21	0

$$LS(e) = LF(e) - t$$
$$LS(e) = 17 - 6 = 11$$
$$LS(f) = LF(f) - t$$
$$LS(f) = 17 - 8 = 9$$

To obtain the LOT for a node we must obtain the LS for all activities leaving the node. The minimum activity LS value will define the value at which the node is realized (LOT) and, in turn, the LF for all activities entering the node. For node 4 the $LS(e) = 11$ and the LOT for 4 is 11 time units. In the case of node 3:

$$LS(d) = 11 - 2 = 9 \text{ and } LS(f) = 9.$$

Therefore

$$LF(b) = 9$$

since

$$LF(b) = \min \{LS(d); LS(f)\} = \min \{9, 9\}$$

The LOT for node 3 is 9.

$$LS(b) = LF(b) - t$$
$$LS(b) = 9 - 6 = 3$$

$$LS(a) = LF(a) - t$$
$$LS(a) = 4 - 4 = 0$$

Therefore, the node 1 LOT becomes $\min \{LS(b), LS(a)\} = \min \{3, 0\} = 0$

Table 7.3 and table 7.4 illustrate the results of the calculations made. Note that many of the calculations can be done within the tables.

Table 7.4 CPM calculations for site preparation project events

Event (node)	EOT	LOT	Slack
1	0	0	0
2	4	4	0
3	6	9	3
4	11	11	0
5	17	17	0
6	21	21	0

The results of the forward and backward calculations allow the project manager to gain an understanding of the schedule with the activity duration times originally assigned. Any changes made to the activity times along the critical path affect the project duration times. For activities outside the critical path, changes in duration and start time can be made when certain conditions exist. If an activity is to be extended in time, then the activities to be changed must have slack time associated with them or the overall project duration will change. The activity slack time identifies the range of time that is available to start an activity without delaying the completion date of the project. The slack time for activities can be found by taking the difference between the latest start time, LS, for an activity and the earliest start time, ES, for an activity. Likewise, the event slack (see table 7.4) provides the project manager with an understanding of how much time the realization of an event can be delayed without affecting the project completion time. Note that activities on a critical path do not have any slack associated with them. Any changes to activities on a critical path will change the project completion time.

The Program Evaluation and Review Technique (PERT)

In principle the Program Evaluation and Review Technique (PERT) and the Critical Path Method (CPM) are very similar. The overall objective of both of these approaches is to derive a workable schedule for a project. Once an estimate of each task time has been established, the basic PERT procedure for estimating the times for milestone points in the project network is the same as the CPM approach. The forward-pass calculations and the backward-pass calculations are the same for PERT and CPM.

A fundamental difference between PERT and CPM is the approach used for estimating the times for the project tasks. The CPM method assumes that the project manager's single estimate of task times is good enough for estimating the project schedule. PERT takes the calculation of project completion times to a higher level of mathematical sophistication. The PERT techniques of analysis use the theories of probability and statistics to evaluate project uncertainties. Since this book does not require the reader to have a background in the mathematics of probability and statistics, we will not attempt a detailed discussion of the many features of PERT. The reader interested in mathematically based scheduling techniques should refer to El-sayed and Boucher (1985) or Meredith and Mantel (1995).

In PERT the task times are calculated using a formula based on three time estimates provided by the project manager or the project team. The three required estimates of the time to complete the project are the most optimistic estimate, the most pessimistic estimate, and the most likely time estimate. The early developers of the PERT technique found that by making some assumptions regarding the probability distribution associated with predicting the task times, a good single value estimate could be made using these three estimates. The benefit of this approach is that the derived estimating procedure results in a task time estimate that takes uncertainty into account.

Figure 7.9 illustrates the concept of three time estimates for a task time. The most optimistic task time (a) might be described as a duration that would occur with a frequency of one

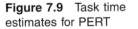

Figure 7.9 Task time estimates for PERT

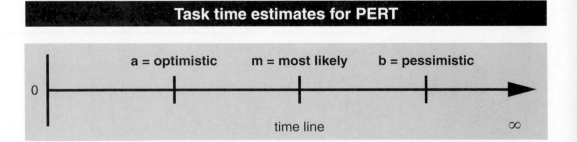

in one hundred. The most pessimistic time (b) might also be a time with a duration that would occur one time in one hundred. The most likely time (m) is the estimated time that the task will most often take. Figure 7.9 demonstrates that the optimistic time "a" is less than time "m," the most likely time. Thus "m" is less than "b."

The formula that the early PERT researchers developed for calculating the single time estimate (T_e) for the task is given as:

T_e = PERT time estimate for a task

$$T_e = \frac{(a + 4m + b)}{6}$$

where

a = most optimistic task time estimate

b = most pessimistic task time estimate

m = most likely task time estimate

See Meredith and Mantel (1995).

Another statistical measure that is an indicator of the variability of the estimate made using T_e is the variance (Var), which can be calculated as follows:

$$Var = \left(\frac{b - a}{6}\right)^2$$

Figure 7.10 presents a hypothetical network with associated task times. The next step in arriving at an estimate for the project schedule would be to perform the forward-pass and the backward-pass calculations using the T_e values calculated for each task as the estimates of task times.

Figure 7.10 PERT network with time estimates

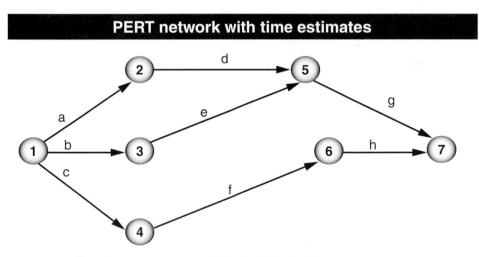

PERT network with time estimates

activity	time estimates			T_e
	a	m	b	
a	1	2	3	2.00
b	2	3	6	3.33
c	4	5	6	5.00
d	3	3	3	3.00
e	3	4	8	4.50
f	1	2	3	2.00
g	4	5	6	5.00
h	3	4	7	4.33

SECTION 7.4   REFINING AND ADJUSTING THE SCHEDULE

The CPM and PERT methods of project analysis can give a project manager considerable insight into the project tasks' start and complete times. However, there is more to scheduling than calculating milestones. The project Gantt chart, responsibility chart, and network diagram provide the project manager with the tools to think beyond just getting the project done and to focus on finishing the project efficiently and according to specifications. The CPM analysis shows how the project can be completed in the time specified by the critical path if task times remain as they were estimated. If the estimated project completion time is later than the time desired by the customer, adjustments must be made.

The project completion date can only be advanced by reducing the duration of one or more of the tasks that make up the critical path. This can be done by bringing in new people, equipment, or other resources. Shortening the duration of an activity that is not on the critical path will not reduce the overall time it takes to complete the project. Reducing the critical path time is tricky and must be done very systematically. Each task has to be examined to determine how much each unit of time reduction will cost in terms of the resources required to achieve the lower time value. The project manager must be aware of the relationship between the time it takes to perform a task and its cost.

When a project time must be reduced, the project manager should select the task on the critical path that will cost the least to shorten and then reduce that task by one time unit. Since the overall project duration will be one time unit less as a result of the task reduction process, it is possible that one or more other paths on the network have become critical. If a new critical path has emerged, the tasks on that new path must also be considered in the time reduction process. The overall project duration can be iteratively reduced one task at a time until either customer requirements are met or the tasks have reached their limits of reduction. Every task has a **limit of reduction,** a point at which the addition of one more worker or another piece of equipment has no positive effect on the task completion time. To successfully reduce the critical path, the project manager must evaluate every unit of time reduction with respect to cost, new critical paths, personnel needs, and equipment needs.

During a CPM analysis, very little thought is usually given to the skill levels required by the workers involved. It is during the post-CPM analysis that the project manager must perform a skills assessment. A problem arises when more of the same type of skilled worker is needed than the system can realistically provide. When too many of a particular class of worker are needed, the work schedule must be adjusted to accommodate the availability of workers. This workload leveling is an important part of the scheduling process.

The Gantt chart is a good tool for assessing the number and type of personnel needed for a project. At any point on the time line of the Gantt chart, draw a vertical line extending upward through the project tasks. For each task that the line passes through, determine the type of workers required to do the work. A tally of the required number of workers provides a count of the people involved with the project at that point in time.

Adjustments and refinements made to the project schedule can be very costly if they are not properly assessed and conscientiously implemented. Rushing through a schedule evaluation can be very costly in the long run if all of the significant variables are not considered.

SECTION 7.5   COSTS AND BUDGETS

A major concern on every project is the ability to control time and costs, because time, money, and resources are directly related. Here are some of the issues that affect costs:

1. The *amount* of time that is charged to the entire effort by the workers
2. The *duration* of time from start to completion

3. The *time* lost due to
   a. factors that are within the control of management
   b. factors that are beyond the control of management
4. The *resources* required to perform the project

Costs and Costing

During the beginning phases of a project, the project manager often prepares initial estimates for the time and costs needed to complete a given item within a project. The assumptions involved in the estimate are noted.

An estimate, which is an educated guess, usually takes into account additional time and money that may be needed to cover unexpected delays, worker inefficiency, and other factors beyond the control of management. Including such amounts in estimates is the responsibility of management.

As noted previously, time and money are directly related. However, never assume that the cost of a task is reduced if that task is completed sooner. Also, it costs less to do a task right the first time than to do it over again. All those involved in the project should continually search for the most efficient amount of time that requires the least expenditure of money.

When complicated projects and programs are initially devised, managers and designers plan for some modest changes. It should be noted that specific changes may not be known at the time that planning occurs. The planners hope that these modest changes will not significantly affect tasks, schedules, or costs, and that the client will observe no overall change in the project plan or performance.

Changes may irritate higher-level managers and clients. They are often more interested in controlling schedules and costs than in knowing how well the work is performed. It is the responsibility of the team to perform quality work, as well as to adhere to the allotted tasks, schedules, and budgets.

The activity center project provides an excellent example of the interaction of time and costs. During the Study Phase, the activity center size and content were determined. When more detailed information was obtained during the Design Phase, changes in the overall plan were made. If the type of lumber originally selected was no longer readily available, its lack of availability would delay the project schedule and would cause the activity center costs to increase beyond budget. Therefore, the related portion of the specifications would be changed via a Contract Change Notice (appendix E), and the schedule would also be changed to reflect the new design.

Costs are normally subdivided into a variety of categories. Two of these categories are **fixed costs** and **variable costs.**

1. Fixed costs include rent, loan payments, and property taxes. The exact amount of these costs can be estimated in advance for a year or more.

2. Variable costs include labor and employer-paid social security taxes, utilities, travel expenses, telephone calls, freight charges, and sales taxes. Costs can also vary when the amount of work contracted or the effort applied to obtain work (Sales and Marketing) changes from the original estimate. Such costs can also be controlled by applying limits (ceilings) beyond which managerial permission must be obtained for further expenditures.

For estimating purposes, these costs are usually divided into different categories, such as the following:

Direct Costs: Labor (salary and hourly) and Materials

Overhead Costs and Rates: Indirect Costs, Facilities, Taxes and Insurance

The overhead costs can be subdivided as follows:

Indirect Costs: Benefits, purchasing, sales and marketing, management, travel, entertainment, education and training (in-house, conference attendance, and courses at schools, colleges, and universities)

Figure 7.11 Break-even
analysis graph

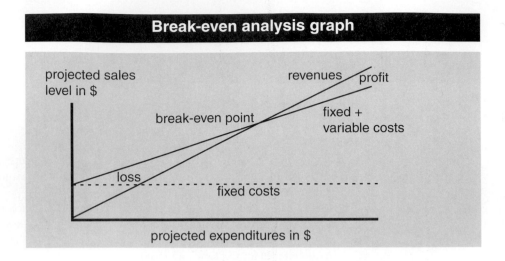

Break-even analysis graph

Facilities: Rent, lease, or ownership; building maintenance and grounds maintenance

Taxes and Insurance: Federal (social security, personal income, corporate income, and unemployment); state (unemployment and workers compensation); local (property taxes and city sales taxes)

These categories will also appear on the appropriate spreadsheets.

Accountants who wish to estimate corporate costs for one or more fiscal years prefer to use computer programs that convert the above categories into fixed and variable costs. Accountants who are analyzing and projecting costs for one or more fiscal years also develop charts, such as the one shown in figure 7.11, that present a pictorial description of these relationships.

The lines on figure 7.11 represent an accountant's summary of sales versus expenditures. The fixed costs, described earlier, are shown as a horizontal line because they change minimally during a given fiscal year. The variable costs increase as more money is spent; they are related to the amount of money spent for developing projects. The **break-even point** indicates where the corporate income changes from a loss to a profit as the income from sales exceeds the costs incurred.

The comparison of project income versus project expenditures is referred to by business and industry executives as the **bottom line.** Executives are judged by how much profit is derived from a project—the amount the income exceeds the expenditures. These executives are judged less by whether the project is performed on time and to the exact specifications. No matter how well a project is performed and no matter how fast it is completed, if the project expenditures exceed costs, that fact is most remembered by corporate executives.

Sometimes a project is stopped (killed) because it becomes clear that the project is going to exceed its originally estimated costs. If additional money is not available, then all the effort and money expended to this point will have been wasted. Most project financial failures can be avoided if costs are estimated carefully at the beginning, during the Conception Phase, and fine-tuned during the Study and Design Phases. Costs need to be monitored and carefully controlled while the actual work is in progress, during the Study, Design, and Implementation Phases.

Budgets

A budget is a plan that describes the authorized expenses for a specified period of time. The word *budget* used as a verb means "to prepare a plan of expected income and expenditures." A budget is based upon an estimate of the income expected so that related expenses can be projected or controlled and remain within a preagreed limit or ceiling (Carrithers and Weinwurm, 1967).

Budget estimates, developed by a technical group within a corporation, are prepared with the assistance of Purchasing Department personnel. The technical people describe the items desired, often with an approximate price included. The purchasing personnel negotiate with vendors to obtain the best price for those items.

Technical personnel are usually required to estimate only the amount of labor required (such as hours or days by labor category) and the approximate cost of materials needed to perform the anticipated project. Accounting personnel then extend those estimates by including the applicable dollar amounts and overhead costs. These two sets of budget estimates are then reviewed by management personnel, who may require that the estimates be revised. It is the responsibility of the technical personnel to indicate when downward revisions would harm the quality of work, or require a change in the specifications.

The budget categories described in the previous Costs and Costing section of this chapter are the ones that are to be used in the cost-estimating effort. Cost estimating for a project is best performed by the designers who have decided on the labor categories to be involved and who have selected the materials needed. If the project is large, then some technical personnel will be assigned to assist the Purchasing personnel in the materials estimation process.

Once the design direction has been selected, it may be worthwhile to compare both labor and materials estimates. If labor category estimates can be combined, then efficiencies result. For example, where needed skills are similar on several tasks, fewer employees may be needed to work on the project, thus reducing administrative and management costs. If common materials for several parts of the project are selected, lower prices can result from buying larger quantities. These efficiencies are best achieved by those designers who have developed the design being estimated. Thus, looking for efficient solutions requires a team approach.

The selection of vendors is critical. Costs can increase if vendors are not dependable in delivering ordered items or if the quality of delivered items is poor. Also the decision to construct items, rather than purchase preassembled items, is critical. This decision involves comparing the costs of labor, overhead, and materials for constructing an item to the total cost of a purchased preassembled item.

The activity center project provides an example of a make-or-buy decision concerning a set of steps. Refer to chapter 4 and figure 7.12.

Make: Students design the stairs, purchase the lumber, and construct the stairs.

Buy: Students (school) purchase a prefabricated set of stairs from a lumber supplier.

The totality of labor and materials costs, convenience, and availability must be compared in a make-or-buy decision. If the cost differential is small, it may not be worth the effort for the company to build the set of stairs. Also, the availability of personnel (when it is time to construct the stairs) is crucial. It is important not to invest more time, and therefore, dollars reaching a decision than will be saved by purchasing the set of stairs from a lumber supplier.

Figure 7.12 Steps comparing make-versus-buy budgets

Steps comparing make-versus-buy budgets			
make		**buy**	
design stairs	$200	select standard design	$100
purchase lumber	$100	purchase prefabricated stairs	$350
construct stairs	$400	install stairs	$100
totals	$700	totals	$550

The estimates versus actuals spreadsheets are used to monitor the financial progress of projects as well as corporations. They can be updated weekly, monthly, or quarterly. The choice of updating periods depends upon how rapidly projected costs are expected to change and how significantly these costs are affecting the bottom line of the corporation. A typical estimates versus actuals spreadsheet is described in section 6.5 and shown in figure 6.6.

As indicated previously, charge numbers can be used to coordinate tasks, schedules, costs, and budgets. A number or coding system—such as a system that is a combination of letters and numbers (alphanumeric)—can be used for project planning, monitoring, and control (Samaris and Czerwinski, 1971).

An example of a numbering system is given below:

XX,YY,6.2.1

This simplified example of a numbering system consists, from left to right, of two letters followed by a comma (XX,), two more letters also followed by a comma (YY,), and numbers separated by periods (6.2.1), where

XX identifies the alphanumerics as either a bid effort or an awarded contract

YY identifies the type of document, such as

ET for estimated task

ES for estimated schedule

EC for estimated costs

EB for estimated budget

AT for actual tasks

AS for actual schedule

AC for actual costs

6.2.1 identifies the related tasks, schedules, costs, and budgets

The number sets are separated by periods. This allows a large project or program to increase the number of digits for a given set of tasks without having to modify previously numbered items.

The numbering system can, for example, be used to identify computer files that contain the related information, such as the document for actual costs. The use of such an alphanumeric system allows those involved to do the following:

- Quickly access task, schedule, cost, and budget documents
- Modify these documents as necessary

Computer programs can "link" documents so that changes made to one document will result in the same changes automatically occurring in related documents. If an individual is not authorized to change a document, that person is notified by a computer-displayed warning. In this way, the data are protected from unauthorized changes.

A trial document is a document devised to resolve an uncertainty. Computers permit rapid preparation of trial task, schedule, and budget documents for management examination and decision making. These trial documents can be manipulated without disturbing the original estimates of tasks, schedules, and budgets. When trial documents are accepted and approved, they can become the applicable set of documents by merely changing their code to replace the original documents.

Sometimes information stored from previously completed work will be similar to that being bid. Data available in computer storage can be accessed to save the time and effort of preparing new estimates. However, it is necessary to verify the labor, overhead, and materials portions of the costs because they may have changed since the earlier effort was completed. Once these verifications and modifications are done, the code on the document may be changed to automatically insert that document into the work being bid. It now becomes an integral part of the proposal.

Tasks, schedules, costs, and budgets are closely related. They are a key part of project planning; they become an important part of project control once implementation is initiated. Experience and teamwork in establishing them and monitoring them are invaluable.

SECTION 7.6   MONITORING AND CONTROLLING A PROJECT

To effectively control a project, the project manager must monitor each task and compare its actual results to the expected results. A good method for monitoring a project is to plot actual progress on the Gantt chart for the entire project. Color coding is very helpful for making comparisons. For example, black blocks can illustrate the planned activity, while green blocks illustrate actual progress. To understand how project resources are being used, the project manager must calculate the difference between planned or budgeted expenditures and the actual expenditures. Cost overruns need to be identified early in the project.

When the progress or costs do not agree with the expectations (plans), it is the manager's responsibility to provide guidance and support in terms of advice and resources. How much action to take is a tough management decision. It depends upon the type of project and the size of the problem. A "loose" style of project management can result in a project getting so much out of control that it fails due to cost overruns and perhaps missed deadlines, also. At the other extreme, overreaction can bring a project to a standstill, and controlling a project too "tightly" makes team members nervous. They then tend to follow the project requirements and budgets so closely that they do not try new ideas or attempt to be creative. This situation is the opposite of what any good project manager wants to accomplish. Monitoring and controlling a project must be done very carefully.

SECTION 7.7   SOFTWARE SUPPORT FOR PROJECT MANAGEMENT

Planning, monitoring, and controlling even a reasonably sized project requires the organization of large amounts of data. For the small projects used to illustrate the concepts throughout this book, the data can be easily organized. However, in the real world most projects consist of more than forty tasks. Each task has to be monitored, and comparisons must be made between the planned time and cost and the actual performance of the task. Reports have to be written; the progress of the project requires documentation.

The project manager must have data available for decision making on a timely basis. When the customer calls and asks for the status on a certain phase of a project, the project manager needs accurate data to quickly prepare a satisfactory answer.

The project-management data manipulation and calculations required can be prepared by available computer software. For this reason, several good project management software packages have been developed. Today's desktop computers use software that was designed for personal computers, and they can handle very large projects that contain large amounts of data to be processed. Surveys by some private marketing groups and a professional society indicate that Microsoft Project is a very popular project management software package; it is applied by many project managers. Microsoft Project provides the tools to graphically display a project via a Gantt chart or PERT format. Microsoft Project also offers the project manager a good report-writing capability. To illustrate the usefulness of a software package for planning, monitoring, and controlling a project, the application of Microsoft Project will be demonstrated in this section.

Before we provide an illustration of the application of software to the challenges of project management, it is important for the reader to understand that software is only as good as the data provided by the project manager. The use of software to document and support the project-management function can save the project manager many hours of analysis. However, the manager must be careful to continually check the reasonableness of every calculation performed

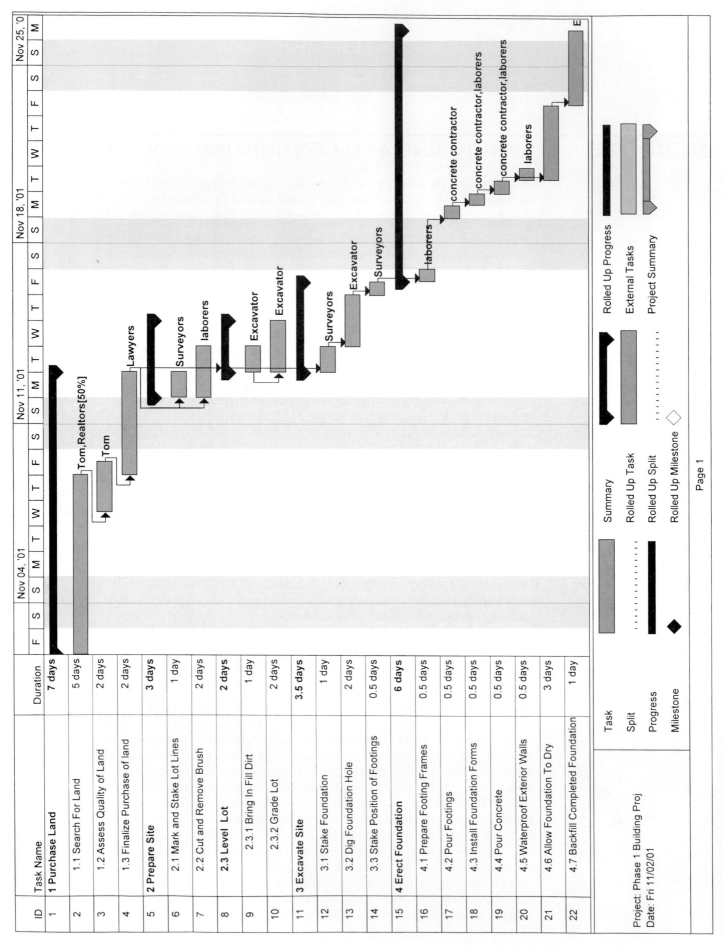

Figure 7.12   Gantt chart from Microsoft Project

Phase 1 Building Proj							
ID	❶	Resource Name	Initials	Group	Max. Units	Std. Rate	Ovt. Rate	Cost/Use
1		Surveyor	S		100%	$50.00/hr	$70.00/hr	$0.00
2		Realtors	R		100%	$0.00/hr	$0.00/hr	$0.00
3		Lawyers	L		100%	$100.00/hr	$200.00/hr	$0.00
4		Surveyors	S		100%	$80.00/hr	$100.00/hr	$0.00
5		laborers	l		100%	$25.00/hr	$30.00/hr	$0.00
6	◇	**Excavator**	E		**100%**	**$100.00/hr**	**$200.00/hr**	**$0.00**
7		concrete contractor	c		100%	$100.00/hr	$100.00/hr	$0.00

Figure 7.14   Phase I building project resource data

by the software. A simple human error, such as the transposition of two numbers, can cause erroneous results that can destroy a project. The software will accept any data provided and it will perform the designated calculations whether the data are correct or incorrect.

The greatest benefit of using a software package (such as Microsoft Project) is the ability to manipulate, modify, and graphically display project schedules. Figure 7.13 illustrates the results of entering the data from the Student Activity Center project, shown in figure 7.3, into Microsoft Project. Data regarding each task on the Work Breakdown Structure can be entered into Microsoft Project using data templates. Figure 7.14 illustrates the set of data stored in the student project database regarding the resources used to accomplish the project. The data must be entered by the project manager or by a knowledgeable assistant. Once the project manager provides the project database with a work breakdown structure, task precedence relationships, duration of tasks, resources required and the cost of those resources, and any fixed dates the project must satisfy, most commercially available software can generate a full range of reports. Figure 7.13 is a Gantt chart for the Phase I Student Activity Center project prepared via Microsoft Project. All of the resources used are shown as labels on the bars of the chart. The precedence relationships that exist between activities are shown by arrows on the chart.

Within Microsoft Project is the built-in capability to prepare a wide range of reports and charts. Figure 7.15 is an illustration of a resource utilization chart. This chart outlines, for all resources used on the project, the number of time units each resource will apply to the project and to which tasks the named resource will be assigned.

In figure 7.16 there is a demonstration of how Microsoft Project can overlay a project schedule on a calendar. Further, figure 7.17 is a Gantt chart in which the Critical Path has been identified and coded. It should be noted that the graphics applied by a particular software package may require some analysis on the part of the user.

Microsoft Project and other software packages offer capabilities beyond the ones illustrated in this section. In many of the packages, the features allow a project manager to devise a PERT chart, to perform resource leveling, and to track a project as it is being completed. For large projects, the project manager must rely on software for planning, monitoring, and controlling the work.

CHAPTER OBJECTIVES AND SUMMARY

Now that you have finished this chapter, you should be able to:

1. Explain the importance of project planning.
2. Describe and develop a Work Breakdown Structure.
3. Create a Gantt chart for a project.
4. Establish who is responsible for project tasks and create a responsibility matrix.
5. Build a network diagram for a project.
6. Analyze the network diagram and establish an acceptable schedule of the work.
7. List the various kinds of costs that must be monitored in a budget.
8. Understand the schedule adjustment process and use software to make the adjustments.

Phase 1 Building Proj

ID		Resource Name	Work	Details	Nov 04, '01									Nov 11, '01			
					F	S	S	M	T	W	T	F	S	S	M	T	W
		Unassigned	0 hrs	Work													
		Allow Foundatio	0 hrs	Work													
1		Surveyor	36 hrs	Work	8h			8h	8h	4h	4h	4h					
		Search For Lan	20 hrs	Work	8h			8h	4h								
		Assess Quality	16 hrs	Work					4h	4h	4h	4h					
2		Realtors	20 hrs	Work	4h			4h	4h	4h	4h						
		Search For Lan	20 hrs	Work	4h			4h	4h	4h	4h						
3		Lawyers	16 hrs	Work							8h	8h					
		Finalize Purcha	16 hrs	Work							8h	8h					
4		Surveyors	20 hrs	Work											8h	8h	4h
		Mark and Stake	8 hrs	Work											8h		
		Stake Foundatio	8 hrs	Work												8h	
		Stake Position o	4 hrs	Work													4h
5		laborers	32 hrs	Work											8h	8h	
		Cut and Remov	16 hrs	Work											8h	8h	
		Prepare Footing	4 hrs	Work													
		Install Foundatio	4 hrs	Work													
		Pour Concrete	4 hrs	Work													
		Waterproof Exte	4 hrs	Work													
6	◇	Excavator	48 hrs	Work											16h	16h	16h
		Bring In Fill Dirt	8 hrs	Work											8h		
		Grade Lot	16 hrs	Work											8h	8h	
		Dig Foundation	16 hrs	Work												8h	8h
		Backfill Comple	8 hrs	Work													8h
7		concrete contractor	12 hrs	Work													
		Pour Footings	4 hrs	Work													
		Install Foundatio	4 hrs	Work													
		Pour Concrete	4 hrs	Work													

Figure 7.15 Resource utilization chart

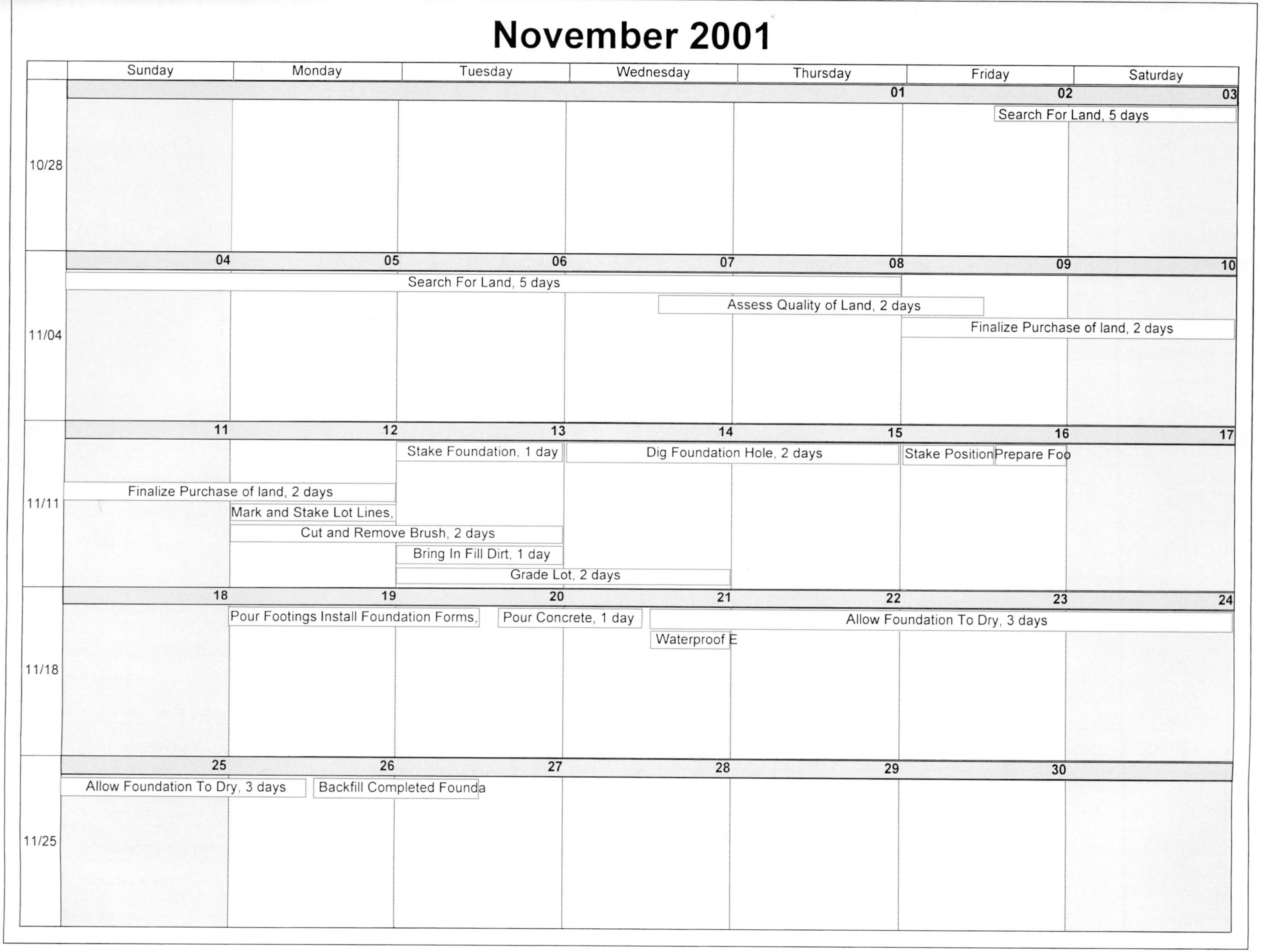

Figure 7.16 Project overlays on calendar

199

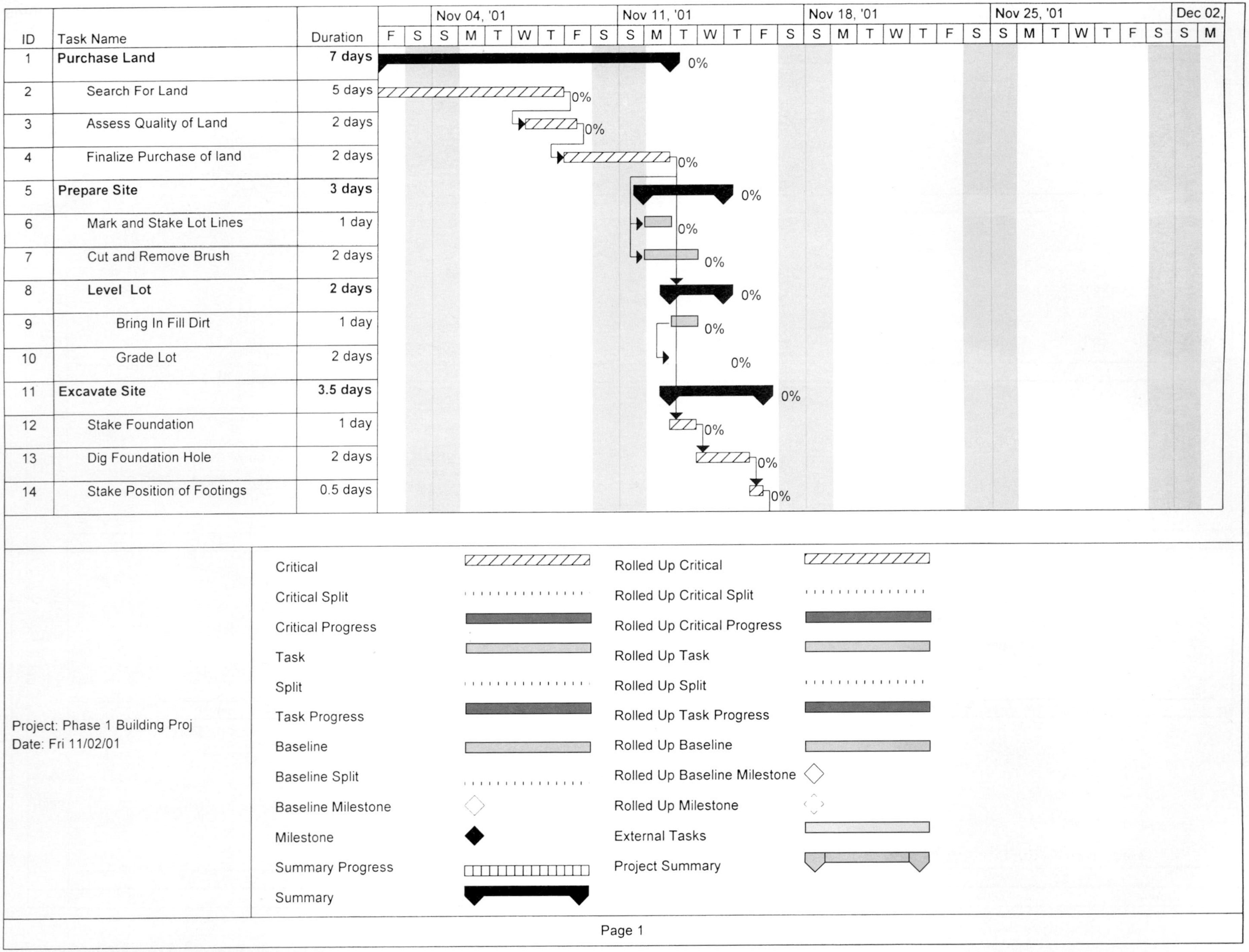
ID
Task Name
Duration
Nov 04, '01
Nov 11, '01
Nov 18, '01
Nov 25, '01
Dec 02,
F S S M T W T F S S M T W T F S S M T W T F S S M T W T F S S M
1 Purchase Land 7 days 0%
2 Search For Land 5 days 0%
3 Assess Quality of Land 2 days 0%
4 Finalize Purchase of land 2 days 0%
5 Prepare Site 3 days 0%
6 Mark and Stake Lot Lines 1 day 0%
7 Cut and Remove Brush 2 days 0%
8 Level Lot 2 days 0%
9 Bring In Fill Dirt 1 day 0%
10 Grade Lot 2 days 0%
11 Excavate Site 3.5 days 0%
12 Stake Foundation 1 day 0%
13 Dig Foundation Hole 2 days 0%
14 Stake Position of Footings 0.5 days 0%
Project: Phase 1 Building Proj
Date: Fri 11/02/01
Critical
Critical Split
Critical Progress
Task
Split
Task Progress
Baseline
Baseline Split
Baseline Milestone
Milestone
Summary Progress
Summary
Rolled Up Critical
Rolled Up Critical Split
Rolled Up Critical Progress
Rolled Up Task
Rolled Up Split
Rolled Up Task Progress
Rolled Up Baseline
Rolled Up Baseline Milestone
Rolled Up Milestone
External Tasks
Project Summary
Page 1

Figure 7.17 Critical Path analysis and coding

EXERCISES

7.1 Write a task description for the work that you are going to do today. Be very specific as to how you will identify the beginning and ending points for the task.

7.2 Preparing a meal can be a very challenging project. Develop a Work Breakdown Structure for the preparation of a specific meal, such as a backyard cookout.

7.3 Prepare the tasks, schedules, and budgets for an instructor-assigned project.

7.4 Develop a network diagram and perform the CPM forward-pass calculations and the backward-pass calculations to arrive at a schedule for the project assigned in exercise 7.3.

7.5 Describe how you would recommend monitoring the progress of a project utilizing the Work Breakdown Structure, the Gantt chart, the responsibility chart, and the network diagram.

7.6 Consider the situation where a member of your family (or yourself) is considering starting a mail-order business from your home. List all of the fixed and variable costs associated with such a venture.

1. How are specifications developed through the phases of a project?

2. Why are trip and meeting reports useful, and what do they typically contain?

3. What are the types of contracts for projects?

4. What are periodic reports for projects? What are their contents? For whom are they intended?

5. What kinds of reports are prepared for government projects, and what are their typical contents?

6. What are the goals of ISO standards?

Specifications and Reports

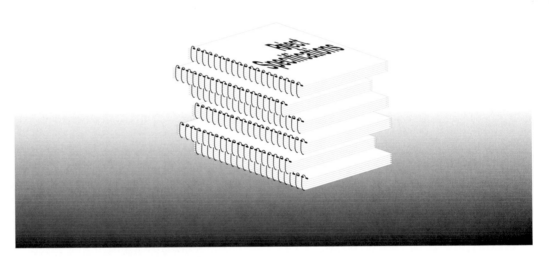

Progress imposes not only new possibilities for the future but new restrictions.

Norbert Weiner (1954)

During the last fifty years, project documentation has evolved into several standard documents, such as specifications, contracts and change notices, meeting reports, periodic project reports, and government reports. Each of these documents, or groups of documents, is discussed in this chapter.

Documentation of the individual project phases has been described at the end of each of the project phase chapters. The present chapter includes a more in-depth description of the contents of project specifications and the way in which specifications change during the Conception, Study, and Design Phases. In the Implementation Phase, specifications become critical to the writing of contracts and change notices.

Specification descriptions become more detailed with each phase. The type of information and level of detail required for meeting and project reports remain fairly constant, however, regardless of the specific phase in which the reports are prepared. Government reports are many and varied. Often, different types of government reports are associated with different phases of a project.

After you complete the study of this chapter, you should know how to do the following:

- Participate in the preparation of a specification.
- Understand the content of contracts and change notices.
- Write meeting reports.
- Contribute to periodic project reports.
- Provide information to be included in appropriate government reports.

SECTION 8.1   PREPARING SPECIFICATIONS

At the beginning of the planning process, an overview of the entire sequence of a project is devised. That overview consists of the following documents:

- Description of the expected final product

- Approximate schedule that leads to the final product

- Estimate of total costs that will lead to the final product

- Initial list of the work that will eventually lead to the desired product

The simplest and most direct way of moving forward from this initial effort is to assign team members to write individual task descriptions representing work that will lead to the final product. These descriptions are trial documents that are circulated among the team members. As they are revised, they become part of the Conception Phase documentation (see Section 2.3). However, it is not possible to write detailed task descriptions without reference to standards of design and workmanship that help to define the final product.

Specifications (specs) are documents that supply those standards. Specifications describe the desired end result of the work to be accomplished. They contain requirements that are the conditions to be fulfilled in order to complete a task or project. In addition, specifications typically describe desired features for the project. The sequence of specification development is given in figure 8.1.

Conception, Study, and Design Phase Specifications

PLANNING AHEAD

During the Conception Phase of a project, a Preliminary Specification is prepared. The content of this specification depends upon the type of projects being considered. Notes are included that identify entire standards or specific paragraphs to be referenced.

During the Study Phase, the Design Specification is written to define both design requirements and design limitations or constraints. Although these specifications place constraints upon the designers, they should *not* define design methods unless the method is a regulatory requirement. The designer needs as much flexibility as possible.

Figure 8.1 Specification
development

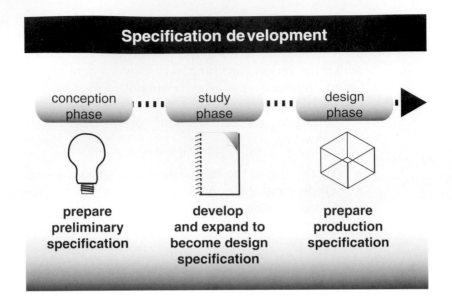

The object of the Design Specification is to stimulate the designer's imagination. An example of a design-stimulation statement is the following:

The selected project is an activity center to be used by children from ages four through ten.

This information is a general guide to the designers.

At the same time, the Design Specification should note any constraints that must be considered. An example of a constraint is

The cost of the activity center shall not exceed $20 000.

Depending upon the available land, constraints include: local building-code restrictions, environmental guidelines, and activity center-site physical conditions.

During the Design Phase, the Design Specification is expanded to include sufficient detail to build or manufacture the product. The resulting Production Specification defines the design requirements and design limitations or constraints. It should not, however, define manufacturing methods. This effort is the responsibility of the manufacturing personnel.

Any questions or options remaining from the Study or Design Phases must be resolved prior to the start of the Implementation (or Production) Phase. Thus the specifications to be used during this last phase of the project are now final.

The object of the Production Specification is to guide the manufacturing (or construction) personnel and, at the same time, note any constraints that must be considered. An example of a construction specification statement that allows for flexibility is the following:

The type of lumber may be spruce, white pine, or fir.

An example of a constraint would be the following:

*The wall and roof sheathing shall be 5/8″ CDX (**X** for exterior-glue) plywood.*

The Production Specification is the defining document for the entire technical portion of a project. This specification must be detailed and precise so that once the project is complete, the finished product will meet the specification. This specification may also contain tests, or references to tests, that will verify product acceptability.

The Production Specification and the documents that are referenced in it form the technical portion of a contract. A precise specification ensures that two or more groups bidding on the same project will be fulfilling the same requirements. This is often referred to as comparing apples to apples, rather than apples to oranges. Where a particular equipment or device is desired, specifications may include a brand-name reference or references along with the phrase "or equivalent." Then it is the responsibility of the bidders to propose an equivalent replacement for that specified item if they so desire and if they can show that the replacement is equal to or better than the brand name(s) referenced.

Specification Content

The organization and content of a typical specification are summarized in figure 8.2. The terms in figure 8.2 are explained in the following list.

Scope of Specification typically contains a Specification Descriptive Title and a brief narrative description of the specification content.

Applicable Documents reference those documents that are to be considered a part of the specification. This listing eliminates the need to repeat all the standard requirements already contained in these documents, which are often referred to as "boilerplate." (The term originates from the boiler design of a steam engine, which requires the use of strong material.) These boilerplate documents have been standard reference documents for many decades.

Requirements consist of many paragraphs that describe the following:

System Definition is a general description that includes the purpose of the specification, a system diagram, interface definitions, and, if government sponsored, the Government Furnished Property List.

System Characteristics describe the system's physical and performance characteristics, which include: reliability, maintainability, availability, system effectiveness models, environmental conditions, and transportability where applicable.

Design and Construction includes: materials, processes, and parts; nameplates and product markings; workmanship; interchangeability; safety; and human performance/human engineering information.

Figure 8.2 Hierarchy of specification content

Hierarchy of specification content

scope of specification

applicable documents

requirements
    system definition
    system characteristics
    design and construction
    documentation
    logistics
    personnel and training
    precedence

quality assurance provisions

preparation for delivery

notes and appendixes

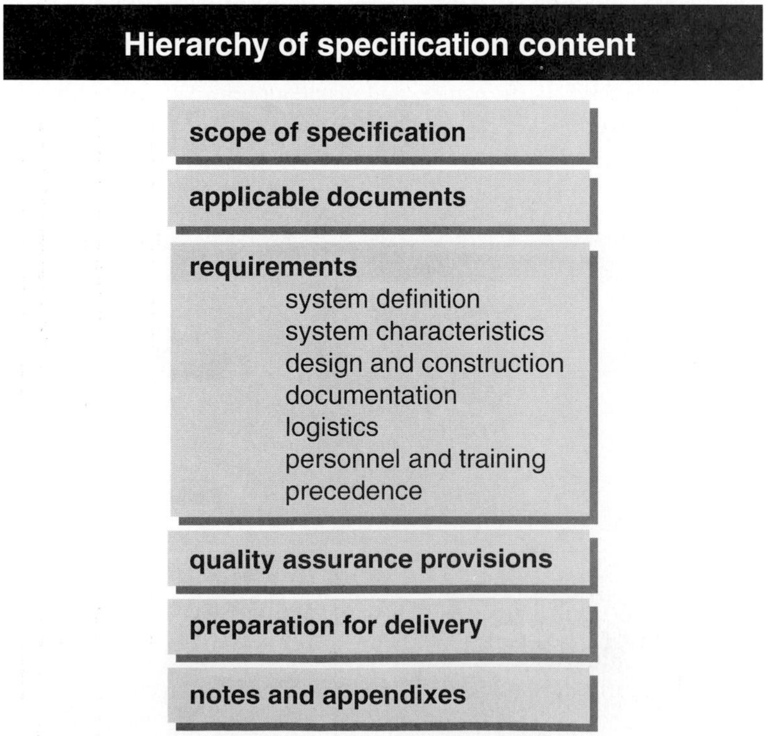

Documentation refers to the complete sets of drawings, instruction manuals, performance and acceptance test reports, and scale models (if they have been constructed).

Logistics lists maintenance and supply facilities, and facility equipment for use with the implemented system.

Personnel and Training describes employee skills, training requirements, training facilities, and instructors with their instructional material necessary to prepare those people who will operate and maintain the system.

Precedence references those documents that will take precedence over another should any two or more referenced documents conflict.

Quality Assurance Provisions include specifications for tests and procedures to verify that the equipment and/or system has been designed and constructed to meet the quality-level requirements.

The "general" paragraphs assign responsibility for all tests. They reference any tests and examinations required by the specification.

Quality Conformance Inspections include on-site inspection of test facilities, test procedures, and test personnel by qualified inspectors appointed by the client or the government.

Preparation for Delivery includes specifications for acceptable packaging and shipping methods.

Notes and Appendixes typically include any comments that apply to the various paragraphs within the specification.

Drawings are used where words are insufficient. Three-dimensional scale models may be constructed. Thus, more complicated devices and systems can be examined visually and understood by nontechnical personnel who may not be capable of interpreting technical drawings.

Levels and Types of Specifications

There are various *levels* of specifications (see figure 8.3). The system-level specification is an overview of the project in terms of its overall performance or function. Design specifications are more detailed and can be further subdivided into component or device, performance, installation (such as National Electrical Code), and environmental specifications.

Maintenance information is often either noted or referenced at the design, device (or subassembly), performance, and installation levels (see section 5.3). For large systems or projects, however, a separate maintenance specification or plan may be prepared. This plan, at a minimum, should contain the following (where applicable):

Figure 8.3 Specification hierarchy

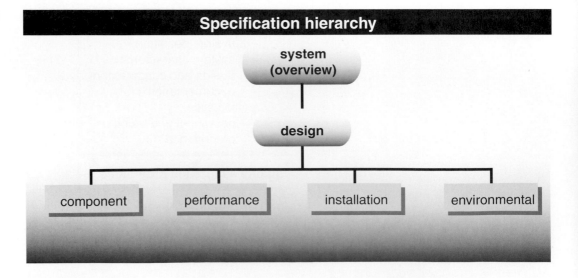

- Class, or level, of maintenance involved
- Suggested schedules for maintenance and parts replacement
- Mechanical parameters, including

> Inspection and cleaning
>
> Testing and lubrication
>
> Measurement locations and tolerance factors
>
> Alignment and adjustment recommendations
>
> Related mechanical, assembly, and part-layout drawings, block diagrams and flowcharts, electrical and/or electronic schematics, and troubleshooting trees

- Electronic test and calibration equipment specifications, including recommended

> Specific corrective maintenance activities, as required performance procedures

Include in the maintenance plan, where appropriate, all precautionary information regarding radio frequency, electromagnetic, or LASER radiation interference. Ensure that instructions never violate OSHA standards for personnel safety during plan activities. Where standards do not exist, follow related MIL-SPEC and MIL-STD practices. These are industry accepted and must be followed for government contracts. If not part of the maintenance plan, include vendor documentation as a separate package along with warranty and other related information.

There are also various *types* of specifications:

- Commercial
- Industrial
- Government

> Local
>
> State (DOR)
>
> Federal
>
>> NIST—National Institute of Standards and Technology
>>
>> GSA—General Services Administration (similar to commercial, and industrial)
>>
>> DOT—Department of Transportation (including the Coast Guard)
>>
>> DoD (Department of Defense—military)—Army, Navy, Air Force, and Marines
>>
>> NASA—National Aeronautics and Space Administration

Applicable codes, often published by national associations, are usually referenced within a specification. For example, the National Electrical Code (NEC) for wiring is published by the National Fire Protection Association. Military specifications (MIL-SPECs) often contain references to military standards (MIL-STDs) for various standard components, devices, and subsystems.

Government specifications are the most complex because they are used for wide and diverse applications. Industrial specifications are usually much simpler because they are directed to more specific applications. The construction industry often follows a format offered by the Construction Specifications Institute (1998).

Specifications are the most important documents of the family of documents described in this chapter. They provide the overall and specific guidelines to the designer and to the management monitoring-and-control team.

SECTION 8.2   CONTRACTS AND CHANGE NOTICES

A contract is an agreement between two or more parties regarding the nature of the product to be delivered, the delivery schedule, and the product cost. See Section 4.3. The contract may be as simple as a Letter of Transmittal or it may be a multipage document prepared by a team of lawyers. Specifications and cost and schedule documents must be referenced within the contract.

There are several types of contractual agreements, such as

- Fixed-price contracts
- Cost-plus-fixed-fee (CPFF) contracts
- Incentive contracts

Each contract type has its unique benefits. These are discussed in more detail in Horgan and Roulston (1989). Typical terms are defined in the appendixes of their book.

A **fixed-price contract** is used where the work to be accomplished can be completely specified in advance. Potential contractors use the supplied documentation to determine the prices for all materials, equipment, and labor. They then prepare and submit a bid and will be paid the prices specified in the contract upon completion of various portions of the project. These payments are known as progress payments. Alternately, the total price of the contract may be withheld until the entire project is satisfactorily completed. If specifications are considered final as a result of careful and complete work during the Design Phase, then a fixed-price contract may be awarded.

A **cost-plus-fixed-fee (CPPF) contract** contains two significant groupings of designs—those designs that will not change and those designs that are incomplete. (The incomplete designs occur because more technical data and experimentation are necessary before the design stabilizes.) The contractor is paid for expenses incurred in performing the work, receiving a previously established fee instead of a percentage of the estimated contract cost at various intervals. A contract for work to be done during the Study Phase or the Design Phase is often a CPFF contract.

An **incentive contract** is one in which the contractor is rewarded with extra money for completing the work within a preagreed time or for less cost than originally estimated. The incentive contract is very valuable to the awarding organization when the contract is one of many interrelated contracts. Failure of one contractor to complete a contract on time may cause very expensive overruns because the other contractors cannot depend upon their start date. Typical contract content is given in the activity center example of appendix D.

Production specifications are seldom perfect. Thus, an Engineering Change Notice (ECN) or Contract Change Notice (CCN) is usually necessary. The wording and content of a typical CCN for the activity center is given in appendix E. The wording and content of a typical ECN is similar.

SECTION 8.3   TRIP AND MEETING REPORTS

When a project is first conceived, it is necessary to gather information and exchange ideas if progress is to be made. Both of these activities usually involve trips and meetings. Visits to libraries, to potential clients, and to other persons or groups must be documented in the form of a Trip Report. In order to cover any type of visit, a simplified form, known as a Contact Report Form, is often used. That form typically contains the following information:

How the contact was made (trip, visit, telephone, or other)

Name and address of the organization visited

Purpose and date of the contact and the person reporting

Name, title, and telephone numbers of persons contacted

Summary of any discussions

How the contact was initiated, the source of the "lead"

Materials collected and where they are filed

Action items and person(s) responsible for taking action

Distribution list of persons who should receive a copy of this Contact Report Form

A sample Contact Report Form is shown in figure 8.4. (The name on the right-hand edge is for hard-copy filing purposes.)

Figure 8.4 Contact Report Form

CONTACT REPORT FORM

Contact was made by: trip □, visit □, telephone □, or _____

Organization Name _____

Address _____ City _____ State _____ Zip _____

Purpose of Contact: _____

_____ Person Reporting: _____ Date _____

Persons Contacted	Title	Telephone
_____	_____	_____
_____	_____	_____
_____	_____	_____
_____	_____	_____

Summary of Discussion: _____

_____

_____

_____

_____

_____

_____

_____

Purpose of Visits: _____

Source of Lead: _____

Materials Collected	Where Filed
_____	_____
_____	_____
_____	_____

Action Items	Person(s) to Take Action
_____	_____
_____	_____
_____	_____
_____	_____
_____	_____
_____	_____

Report Recipients: _____ _____ _____ _____

_____ _____ _____ _____

_____ _____ _____ _____

_____ _____ _____ _____

PROJECT NUMBER

Project progress also requires comparisons; comparisons require examining trade-offs. Ideas are discussed and debated during group meetings of team members, sponsors, and leaders. As shown in figure 8.5, the project team together examines the project concept. The team then studies the various ideas offered and compares them by developing trade-off information. The conclusion is the selected project.

People can spend many hours at meetings arguing about how to proceed and which way to proceed. A **meeting report** records the **minutes** of each meeting. Copies of the minutes are distributed to each participant and to other involved persons. If no one records the discussions and the decision-making process, then the same arguments may be repeated later and meeting time will be wasted. On the positive side, these documented arguments may become valuable resources. If new information becomes available that changes the basis for previous decisions, then these decisions may be modified.

Typical contents for a meeting report include the following:

- Names of meeting attendees
- Topics discussed

Figure 8.5   Project selection

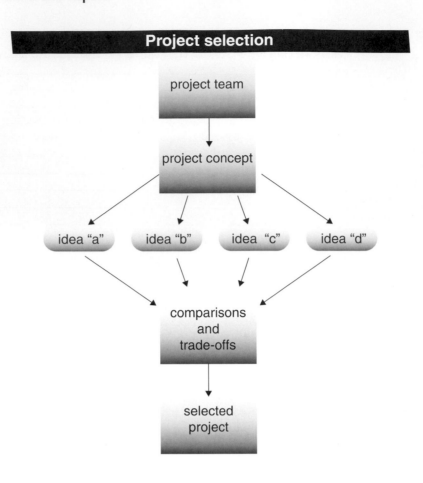

Project selection

project team

project concept

idea "a"   idea "b"   idea "c"   idea "d"

comparisons
and
trade-offs

selected
project

Sample meeting report

Subject: Material Selection Meeting for Activity Center Project   cc: F. Eagle
Date: 1995 June 3   J. James
From: W. Jones   T. Marsh
To: J. Jewett–Manager; Dept 21

Attendees: F. Eagle
              J. James
              W. Jones
              T. Marsh

A meeting was held in Conference Room B to discuss the selection of materials for painting the interior of the Activity Center. It was decided that all ceilings will be painted with one coat of primer and two coats of flat white. The game room walls will be painted with Sunny Peach on the wall above the chair rail, and Desert Sand below the chair rail. The kitchen walls will be painted pale yellow. The colors for the remaining rooms will be decided at our next meeting on June 7.

    Action items
          Purchase 2 gallons of Desert Sand epoxy paint   F. Eagle
          Purchase 1 gallon of Sunny Peach epoxy paint   F. Eagle

Figure 8.6   Sample meeting report

Figure 8.7
Summary of typical
action items

Summary of typical action items

action item	responsible person	completion date
prepare minutes of meeting	W. Jones	June 3
order floor tile material	R. Wells	June 8
select colors and order paint for each room	F. Eagle	June 8
select and order game room equipment	T. Marsh	June 13
install floor material	R. Wells	July 5
complete painting of walls and ceiling	F. Eagle	July 13

- Progress achieved
- Directions chosen
- Decisions agreed upon or made
- Action items to be performed

The document shown in figure 8.6 is considered to be a meeting report.

Action items are very important. They not only describe actions to be taken, they also identify the persons assigned the responsibility for performing the specified action. Thus project progress is ensured. The list of action items in figure 8.7 includes the persons responsible for implementing the items and the required completion dates.

Written words contribute to progress. They can be read when the meeting is over. They can lead to ideas not considered during these meetings. A meeting report is a record of the ideas and progress of the group.

SECTION 8.4   PERIODIC PROJECT REPORTS

Documentation of work on a project includes periodic reports such as daily charges, weekly expenditures, monthly progress reports, and quarterly and annual summary reports. Additional documentation is required at the end of each phase as described in section 3 of chapters 2 through 5. The types of the periodic reports may vary greatly; typical report titles are provided as a list in figure 8.8.

Periodic reports are grouped into three categories:

- Financial reports
- Technical reports
- Managerial reports

Figure 8.8 Typical titles of periodic reports

Typical titles of periodic reports

daily charges
weekly expenditures
monthly progress
quarterly summary
annual summary

Figure 8.9 Managerial
report content and follow-up

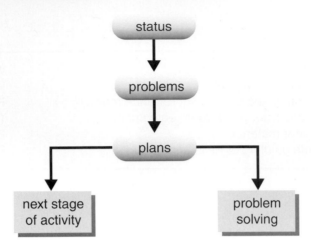

Managerial report content and follow-up

status

↓

problems

↓

plans

next stage
of activity

problem
solving

PLANNING AHEAD ▶

Such reports can be prepared on a computer and are often transmitted over a network to the persons or groups who have an interest in their content.

Financial reports are primarily spreadsheets of detailed data. Where appropriate, this information may be preceded by a summary that compares the original cost estimates to the actual financial data. These spreadsheets contain such items as daily charges and weekly expenditures. Attachments may include explanatory statements where necessary.

Technical reports combine words, diagrams, and drawings. These reports are of greatest interest to the technical manager, whether that manager is part of the design team or part of a Project or Program Office.

Managerial reports combine words, graphics, and summary financial spreadsheets. Managers may not be technical experts and may therefore be more interested in the overall progress of the project and its adherence to schedules and budgets. The content of managerial reports, summarized in figure 8.9, is typically the following:

- *Status:* Work accomplished during the reporting period
- *Problems:* Problems and their proposed solutions
- *Plans:* Work to be accomplished during the next reporting period

When the planning portion of a managerial report is being prepared, an assessment of any risks involved should be included. As noted, managerial reports either include or refer to the applicable financial reports. Often, summaries of those financial reports are included in the managerial reports.

SECTION 8.5   GOVERNMENT REPORTS

Most projects, even those involving only private companies, require some type of government reporting. Each level of government requires the submission of reports or other documents and, where applicable, accompanying money. These levels of government are

1. Local (including towns, cities, counties, or parishes)
2. State
3. Federal
4. International

Documents and financial payments must be prepared and submitted in a timely manner to satisfy government requirements. Typical contents of these reports are described below.

The reports for local town, city, and county—or, in Louisiana, parish—typically include information regarding compliance with building, sanitary, and environmental codes. Other reports indicate the payment of personal property taxes, real estate taxes, and local income taxes, where applicable.

State reports may include information regarding compliance with hazardous waste disposal. They also typically contain payroll, sales, or purchasing information for the payment of state income, sales, unemployment, and other taxes.

Federal reports may include information regarding compliance with OSHA (Occupational Safety and Health Act), ADA (Americans with Disabilities Act), and many other acts that have been passed by Congress. The federal government also has appointed the IRS (Internal Revenue Service) to collect the federal portion of the national unemployment tax, the federal portion of the withheld income tax, and the Social Security and Medicare taxes (FICA and OASDI or FICA-HI).

International reports may include information regarding compliance with standards developed and distributed by ISO (International Standards Organization). Their headquarters in Geneva, Switzerland, is a consortium of 130 contributing countries whose goal is to promote quality standards on a global basis, and to improve the exchange of quality and environmentally acceptable goods and services. These standards contain guidelines prepared by the standards organizations of the contributing countries. The international reports required of companies who sell outside their country are based upon ISO9000 and ISO14000.

Duncan (1996, p. 24) excerpts two ISO definitions that distinguish between a standard and a regulation.

> A **standard** [such as the size of computer disks] is a document developed by a recognized professional body. A standard promotes common and repeated use by providing rules, guidelines, or characteristics for products, processes, or services. Compliance with a standard is not mandatory. A **regulation** [such as a building code] is a document that lays down product, process, or service characteristics including the applicable administrative provisions. Compliance with a regulation is mandatory.

ISO9001 and ISO9002 focus on manufacturing and quality control. Companies must prove they employ an effective quality management system to attain product uniformity and predictability. Any company that manufactures a product falling within a specified industry category must certify that its quality system meets the appropriate ISO standard. Such certification reduces or eliminates customer quality audits. This compliance requirement is part of the product safety laws of many European countries and is being considered for inclusion in other countries around the world. ISO14000 is a similar series of international standards that concentrate on the protection, preservation, improvement, and reconstruction of the environment. To comply with this standard, products must meet the following requirements.

- Material selection should be based on the following criteria to reduce waste:

    Minimize toxic waste content

    Incorporate recycled and recyclable materials

    Use more durable materials

    Reduce total material use

- Production standards and procedures should have the following effects:

    Reduce process waste

    Reduce energy consumption

    Reduce use of toxic materials

- Designs for recycling and reuse should include these features:

    Incorporate recyclable materials

    Disassemble easily

    Reduce materials diversity

    Include labeled parts

    Use standard material types

- Designs that extend the useful life of products and components should include the following features:

   Design for remanufacture

   Design to permit upgrades

   Designs that make maintenance and repair techniques and procedures easier

   Designs that incorporate reconditioned parts and assemblies
- Designs should describe end-of-life safe-disposal techniques.

There are two aspects of the ISO14000 series. The first aspect is the organizational element that addresses environmental management systems. The second aspect addresses product-oriented issues. ISO14001 describes environmental management systems; ISO14010 is the guideline for environmental auditing.

CHAPTER OBJECTIVES SUMMARY

Now that you have finished this chapter, you should be able to:

1. Describe how specifications are developed through the phases of a project.
2. Describe the types of contracts for projects.
3. Explain the importance of trip and meeting reports, and describe the typical contents of these reports.
4. Name the categories of periodic reports and describe their contents. Identify who receives them.
5. Name the various levels of government for which reports are prepared. Describe the typical contents of these reports.
6. Describe the goals of ISO standards.

EXERCISES

8.1   Write a meeting report based upon a group of your peers who have worked together and have met to describe that work. Identify the action items that are to be accomplished before they meet again.

8.2   Develop a change notice form that would be applicable to the work with which you are involved.

8.3   Write a project report describing your monthly accomplishments for either a project or for this course.

8.4   Develop the outline of a specification for a project to which you have been assigned that describes and constrains that project. Indicate how that specification could be modified to become the final specification.

8.5   From your previous work, develop a set of documents that describes your work. Describe how it could be presented to future employees so they can improve their contributions based upon your experiences.

chapter 9

Project Management

The Project Plan

Specifications and Reports

Modeling and System Design

Project Management

The Project Plan

Specifications and Reports

Modeling and System Design ▶

The Need for Models

Human Factors Considerations

To Model or Not to Model

Modeling Applications

Block Diagrams

Expanding the Modeling

Model Interconnecting and Testing

Modeling Throughout the Project Phases

1. What are models, and why are they used in the development of designs?

2. What kinds of models are typically used in the development of designs, and what are examples of each?

3. Why is it important to decide early in a project what to model?

4. What criteria should a model meet?

5. What are modules, and why is it advantageous for models to be designed as modules?

6. What are block diagrams and flow diagrams, and how can they be useful in preparing models for projects?

7. What kinds of activities are performed in relation to models during the four phases of a project?

8. What is the study of human factors, and why is it important in design?

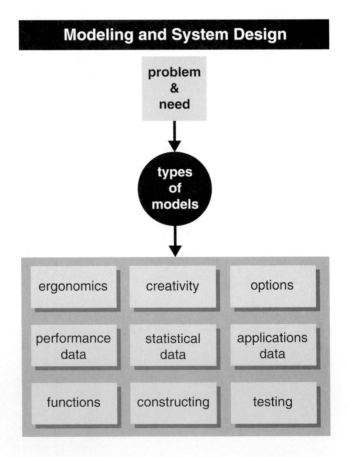

What is false in the science of facts may be true in the science of values.

George Santayana, 1900

Simulation models may be used to replicate a portion of a proposed or existing system. They may also be used to replicate an entire system. The selection of appropriate models for a project occurs as part of the system development process.

This chapter focuses on the values and uses of modeling as related to a system design. Simulation models may be used to replicate a portion of a proposed or existing system. They may even be used to replicate an entire system.

After you complete the study of this chapter, you should know how to do the following:

- Realize when the use of a model would be advantageous.
- Determine where a simulation would be helpful during design.
- Develop a system block diagram that can be used for a model design.
- Examine the applicability of a model to design needs.
- Understand how various models can be interconnected and tested simultaneously.
- Comprehend how modeling can be applied throughout the phases of a project.
- Learn when to apply human factors techniques and guidelines to either a model or to the actual device or system

Rechtin (1997) defines a model as a representation of selected features of a system. It may be a physical scale model of a proposed structure, it may be a defining set of equations, or it may be an operating prototype of a product.

When do we stop modeling and start building our product? Modeling often lasts for the entire life of the project! The initial model is gradually improved until it closely simulates the final product. Models help identify those characteristics that are not immediately evident to the designers. The decision about how much modeling is needed as a part of the system development process must be considered by all designers. See Reutlinger (1970) and section 9.7 in this chapter.

SECTION 9.1 THE NEED FOR MODELS

Modeling is similar to a miniature project to be performed within the overall project. Models are used when there is a significant concern regarding the feasibility of all or a part of the design approach being considered. Modeling a proposed design should be considered when the cost in labor, time, and materials involved in designing and implementing the model is significantly less than that of constructing the actual product. The usefulness of the model in testing the effectiveness of a design must be proportional to the costs involved.

A model is both an abstraction and a simplification of an event or function in the so-called real world. The purpose of a model is to simulate the actual system—or a portion of that system—in an effective and efficient manner. A model must represent the key system characteristics as closely as possible.

Since the purpose of a model is to simulate the actual system—or a portion of that system—in an effective and efficient manner, a model must represent the key system characteristics as closely as possible.

The model is *not* the actual system, nor a portion of that system. Models are simplifications of events and functions in the real world. However, a model that is too simple may lead to incorrect or incomplete results because it does not accurately reflect the real-world situation. Such results would then guide the designers improperly. Therefore, it is mandatory to evaluate the limitations of a proposed model before using the information gained from the study of the model.

Figure 9.1 Types of models

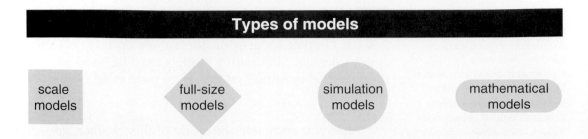

Modeling often occurs during a feasibility study. Studying the operation of a model can lead to ideas that improve a system design. These designs can then be tested before being applied to a project. Some benefits of a model are these:

- Models can demonstrate that a design is unsatisfactory, which can prevent waste of design and implementation costs on products that would not be acceptable.
- Design direction can be obtained from the data gathered from testing the model.

Models can be devised for items as varied as metalworking, insurance policies, chemical processes, and transportation systems. For each area of application, specialized expertise is needed to design a model.

Models can be devised for items as varied as metalworking, insurance policies, chemical processes, and transportation systems. For each area of application, specialized expertise is needed to design a model.

The purpose of a model is to simulate the actual system, or portion of that system, in an effective and efficient manner. There are four types of models (see figure 9.1):

- Scale (reduced-size) models: both static and dynamic
- Full-size models or mock-ups: for technique and process evaluation
- Simulation models: for design, evaluation, and training purposes
- Mathematical models: for test and evaluation via computers

For each type of model, many options exist. Each model typically offers special features, which may be more suited to particular applications. Also, note that a prototype is a type of model. It is a sample of the production version. Prototypes are used to test system and equipment functions. They may not look like the final production version; for example, they might be constructed with different materials.

Here are examples of each type of model:

- Scale model of a radar, for evaluation of radar beam patterns
- Full-size model of the same radar, for evaluation of construction techniques and materials
- Simulation model, for evaluating how many radars are needed to ensure adequate coverage of the specified surface area or volume
- Mathematical models, for analysis of risk (Raferty, 1994) and for data analysis via computer simulations

In addition, design-deficiency monitoring and evaluation models can be used for data gathering and analysis, for example, for studying automotive vehicle safety. Each model should be as modular as possible so it can later be adapted and applied to other design situations.

It is the responsibility of the team leader or manager to decide which modeling approach to apply, for example, weighing the effects of relying totally on a computer simulation rather than constructing a prototype model. The trade-offs include time and money. An inaccurate or poorly designed model may ultimately result in the canceling of the entire project.

During the Conception and Study Phases, models can be developed and manipulated to predict performance outcomes. These outcomes can be used to guide the work of the Design Phase.

Figure 9.2   A typical model

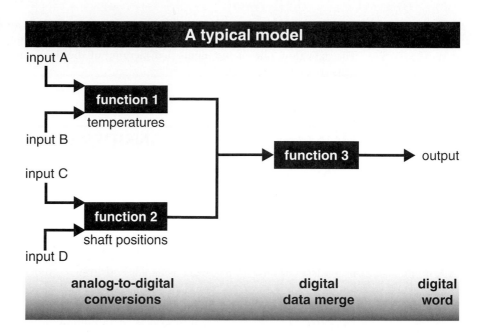

In figure 9.2, Function 1 combines and modifies input A and input B. Function 2 combines and modifies input C and input D. Function 3 modifies inputs from function 1 and function 2. Here is a more specific explanation:

- The inputs to function 1 (input A and input B) might be the outputs of two sensors that are monitoring temperature.

- The inputs to function 2 (input C and input D) might be the outputs of two shaft-to-position transducers.

- Functions 1 and 2 subsystems calibrate the input data and convert these data to digital information.

- Function 3 would then merge or multiplex the digital data into one digital word that will become the output of function 3.

A computer or physical simulation of this subsystem (see figure 9.3) would include the characteristics of a variety of sensors and transducers to ensure that the design works and to

Figure 9.3   A model showing data merging

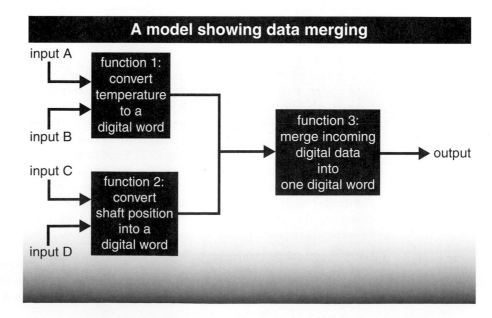

establish the effects of error on the remainder of the system. In this example the temperature sensors, shaft-to-position transducers, and digital-word-merging (multiplexer) are simulated. The data flow rate of the inputs and merging of the data can now be studied without actual construction of equipment.

SECTION 9.2   HUMAN FACTORS CONSIDERATIONS

Most products will interact in some way with people. People may operate them, maintain them, be served by them, or be otherwise influenced by them. A product will be more efficient and safer to use if the characteristics and limitations of the users are considered in its design. The application of knowledge about human beings to the design of products is variously referred to as **human factors** or **ergonomics.**

*A product will be more efficient and safer to use if the characteristics and limitations of the users are considered in its design. The application of knowledge about human beings to the design of products is variously referred to as **human factors** or **ergonomics.***

The physical capabilities and limitations, such as size and strength of the people who will use a design, must always be considered. For lifting or moving heavy objects, people may require training in using the optimum procedures to do so. People may require assistance in the form of devices such as mechanical levers or powered lifts.

The scheduling of work-rest cycles can profoundly affect the performance and well-being of workers. For example, much attention has been given to determining optimum rest and exercise periods in jobs that require continuous operation of computer keyboards.

The operation of human senses needs to be considered in designs. For display information—relating to sense of sight—size, shape, brightness, color, and more legible written characters can all be used to improve transfer of information.

In many applications, the meaning of color codes has been standardized. Here are some examples:

green = safe

amber or yellow = caution or warning

red = danger

white = status (**ON, STANDBY**)

Critical elements should be coded in other ways also, since some people are color-blind.

Aircraft pilots and navigators complain about the color coding used in cars. Imagine the nervousness experienced by these aircraft-cockpit-oriented persons when they are driving a car in which red lights are used as merely indicators, rather than as warnings!

Information may be conveyed to senses other than sight. Vision works only when an operator is looking at a display; audible alarms can attract attention no matter where the operator is looking. Odor-producing substances are added to odorless toxic gases so that the gases can be detected. Attention has also been given to coding items through touch and vibration so that they can be distinguished.

In Western cultures, people learn to read from left to right and from the top to the bottom of a page. Thus, people from these cultures tend to scan computer displays or meters, dials and knobs, the controls, and readouts in the same way. Such items should be designed to exploit this manner of scanning. For example, the most frequently used or most basic information might be located at the upper left. Similarly, controls should match expectations. For example, controls should move to the right or up for **ON** and to the left or down for **OFF.** Moving elements in displays should move in the same direction as the controls that activate them.

Controls must be designed and located to be compatible with the size and strength of expected users. Often, requirements must be developed for the selection and training of personnel. This ensures that they have the skills and knowledge needed for safe and effective operation of products and systems.

The environment needs to be designed so those persons controlling system processes, operating important equipment, or monitoring systems are able to concentrate their attention on their assigned tasks. Distracting noises and colors should be avoided wherever possible. However, the environment must be neither so quiet nor so monotonous that an operator falls asleep.

Considerable attention has been devoted recently to problems arising from long hours spent at computer work stations. These include back and neck problems that might be alleviated by improved design of seats to provide adequate support; carpal tunnel syndrome (arm pains, hand numbness) from inadequate support of wrists during continuous operation of keyboards; and headaches resulting from glare spots and inadequate character size and contrast on display screens.

To the extent that the environment affects people, it will also affect design. Workspace design should consider amount of space, illumination, accessibility of work and maintenance areas, temperature, humidity, noise, and vibration. For example, some manual controls may be inoperable by gloved or cold hands.

These analyses are intended to determine how best to design equipment so that it will accommodate the operators expected to use it. The purpose is to disover how to make the equipment "user friendly", easy to use, and "idiot-proof." For example, a large amount of human factors analysis is applied in the design of aircraft, submarine, and NASA cockpits, not so much to make them comfortable for the pilots, but to make information instantly available to them when they need it. Whatever the specific intent, the general intent is to determine methods that improve human performance.

Specialists who apply knowledge about human behavior to design are known as human factors or ergonomics specialists. They are generally educated in both psychology and engineering and are often additionally educated in anthropometry, physiology, sociology, medicine, or other human-related sciences. Human factors specialists help to design products and systems that can be operated and maintained more accurately and easily and, hence, more efficiently and effectively. These specialists also strive to ensure users' safety and comfort in operating products and systems, as well as to prevent long-term deleterious effects on the user's well-being.

Very large systems or project teams have their own human factors groups. Smaller efforts may employ one or more temporary consultants. Engineers on small projects may have to do their own designing and evaluating. They can refer to textbooks and handbooks that describe human factors or ergonomics. However, they should have a consultant review and critique their efforts.

Representative textbooks in human factors include Dinsmore (1990) and Sanders and McCormick (1992). Two sources for data related to human factors are Karwowski (2000) and Salvendy (1997).

SECTION 9.3   TO MODEL OR NOT TO MODEL

PLANNING AHEAD

During the Conception Phase, areas are identified in the proposed design that may cost too much to implement or that might even cause the design to fail. If such areas are modeled and tested early in a project, the results can be a workable—or even improved—end product.

Examine figure 9.4. The Conception Phase requires planning the Study, Design, and Implementation Phases of a project. At this time, areas within the proposed design that may cost too much to implement or that might even cause the design to fail are identified. If such areas can be modeled and tested and if useful and accurate results are obtained, then the Study and Design Phases—and thus the entire project—will benefit.

During the Conception Phase, any questionable system characteristics, poorly defined variables, unusual design conditions, and vague assumptions need to be identified. The value of using a model is examined early in the project. Why? Modifications at later stages are more costly because they usually affect work already done. In fact, the project may even be delayed as the modifications to the design are studied and implemented. The system specifications must also be revised to reflect the results of the modeling.

When past experience is not considered a sufficient guide to a new design, then feasibility studies should be performed that include modeling. A trade-off must occur between the cost of designing and evaluating an **exact replica** of a portion of the system to be studied versus studying a less expensive **equivalent model.** The word *equivalent* is the key word. The Conception Phase team must continually compare the cost of a more accurate model versus the expected savings in time, money, and design complexity that a less accurate model could offer.

Figure 9.4 Modeling
sequence

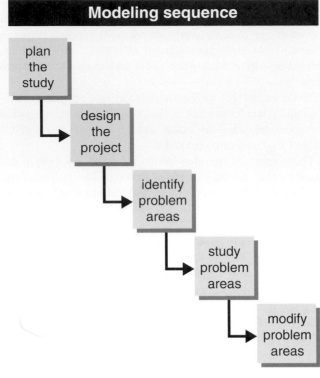

The Conception Phase team
must continually compare the
cost of accurate modeling
versus the expected savings in
time, money, and design complexity that a more accurate
model could offer.

Any model to be considered must be worth the investment of time, money, and designer talent. An appropriate model should meet most of the following criteria:

- It should be relevant to the system design.
- The output from building and testing the model should result in useful and accurate predictions.
- Realistic relations should exist among the various system components.
- There should be a minimum of complexity.
- The format of the model should be flexible and simple to change.
- The model results should be easy to compare with real-world observations.
- Devising and applying the model should cost significantly less than construction of the actual device.
- The result of using the model should lead to cost reductions in system design.
- The result of using the model should lead to improved performance of the system.

Modeling can be applied to research, management, education and training, plant layout, production facilities, equipment design, and equipment operation.

SECTION 9.4   MODELING APPLICATIONS

A desirable feature in models
is that they be modular. Models that are modular can be
tested and validated for the
current project, as well as be
available for use in future
projects.

A company or team that expects to continue its existence beyond the present project will develop models that are modular. Why? The module—or the information obtained from designing the module—can be carried to other projects or systems.

In the field of computer programming, for example, the concept of modularity has led to object-oriented modular programs. These modular programs are designed to be used over and over again as parts of different programs. The modular programs can be used flexibly, and they can be adapted for use in other more extensive and sophisticated programs.

Radar design can only be accurately evaluated by scale models or full-size models. For large airports where a number of radar systems may be operating simultaneously, computer simulations of their interaction continue to be too approximate. Scale models of radar equipment are still used to more accurately simulate expected airport conditions. The data gathered are then compared with measurements taken after the full-size radar equipment is constructed and installed. Then the value and accuracy of the scale models can be evaluated and revised to be more accurate for the next simulation.

In the area of building construction, different types of materials, such as woods, metals, and plastics, are used. Some buildings are constructed, monitored, and evaluated over a number of years. For example, for buildings constructed in desert climates or at shoreline locations, the effect of the climate on different materials can be studied. Data regarding temperature, sun exposure, humidity, salt spray, and other variables are gathered. The materials and construction techniques used are then evaluated for the purpose of determining the most appropriate materials and of developing better construction materials and techniques.

Mathematical models lend themselves to computer implementation and analysis. The continually decreasing cost of computer use allows many designers to consider computer simulation for more rapid and less costly evaluations of proposed designs. Computer simulation can be very flexible and is often quickly adapted to new concepts, such as the following:

- Wind tunnel testing of skyscraper scale models
- Architects' $^1/_{100}$ scale model presentation via models of buildings
- Process plant piping layout
- ASTME-119 standard fire resistance rating tests
- Failure mode and effect analysis (Pugh, 1993, p. 208)

The general limitations regarding computer modeling are these:

- The amount of computer Random Access Memory (RAM) available
- The computational speed of the computer

The continually decreasing cost of computers and their increasing use allows many designers to consider computer simulation for more rapid and less costly evaluations of proposed designs for automobiles and aircraft.

Also, if a computer model is extremely complicated, then it may be very slow and difficult to operate.

Where a design requires repetitive use (figure 9.5), such as a program for analyzing the output of an assembly line that produces semiconductors, the statistics gathered on wafer size, conductivity, response time, and so on, can be presented in a tabular format. The statistical tables can then be activated by the programmer or operator to analyze the performance and trends of the existing production line. Thus, corrective action can be taken before the semiconductor output no longer meets the design specifications.

Automotive vehicles are similarly monitored. The U.S. federal government has developed an evaluation and "recall" procedure. This procedure assures the driving public that noted vehicle

Figure 9.5 Production line data analysis

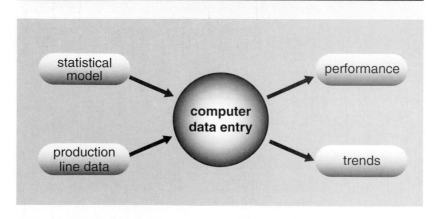

deficiencies are corrected once the problem has been identified and evaluated. Corrective parts and replacement procedures are developed and distributed to local repair stations. The public is then notified to bring in their vehicles for correction. (The federal government has become involved because of the harmful effect of vehicle malfunction on a large portion of society.)

Pugh (1993, p. 208) suggests asking the following general question: What area, if improved, would likely yield the greatest reduction in severity of failure and the frequency of its occurrence, and would also increase failure detectability? Brown (1992, p. 36) offers the following nine specific questions:

1. Can the design be changed to eliminate parts?
2. Can the present design be purchased at lower cost?
3. Can a standard part, or a modified standard part, be used?
4. If the part is to improve appearance, is its use justified?
5. Can the design be changed to simplify it?
6. Can standard inspection equipment be used?
7. Can a part designed for other equipment be used?
8. Can a less expensive, or newly developed, material be used?
9. Can the design be modified to permit manufacture on automated equipment?

Cross (1944, pp. 24–32) warns that the "real" problems must be identified prior to performing a value analysis. It is not always easy to identify the real problems until the design is partially complete.

SECTION 9.5   BLOCK DIAGRAMS

Prior to deciding what portion of a proposed design is to be modeled, the entire design should be translated into a block diagram. Such a diagram defines the input to each block, the functions of the block, and the output from the block. Each block represents a part of the product. The interconnections of the various blocks indicate the relationships among the parts of the product.

Block diagrams are useful in visually dividing a product into its component parts. The blocks can be analyzed to see whether new models should be developed and tested or whether models that have been developed for other projects can be used or adapted.

The Conception Phase team can now ask such questions as these:

- Which blocks within the design are well known and understood?
- Are there blocks that have been used elsewhere in a similar fashion?
- Which blocks consist of newly proposed designs?
- Are the blocks interconnected in a manner previously used?

Based on the answers to these questions, the Conception Phase team may decide that some design issues need to be further explored. Research into existing similar designs may need to be performed.

Examine the block diagram for a bread toaster (figure 9.6) that is to be designed for production. This block diagram shows the designers the functional relationship of each part of the toaster to the other parts. Manufacturing, however, requires a different diagram—one that describes the flow of the toaster's construction and assembly. This flow is illustrated in figure 9.7.

From this flow diagram, the manufacturing functions of purchasing (make-or-buy decisions), tooling, fabrication, assembly, inspection, and testing may be planned.

First, the **ideal system solution** is proposed. For the above manufacturing functions, models are devised. The typical sequence is as follows:

Define the problem

Identify potential solutions

Evaluate alternatives

Select one solution

Figure 9.6 Toaster block diagram

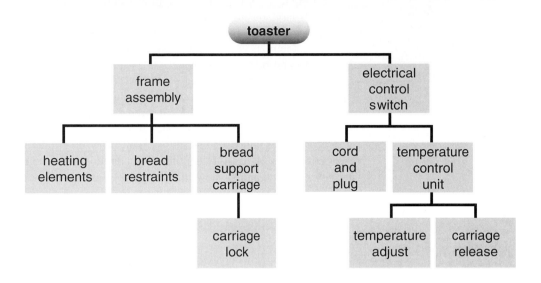

Toaster block diagram

Figure 9.7 Toaster construction and assembly flow diagram

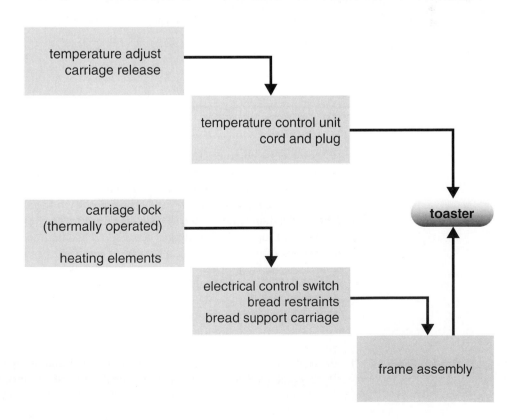

Toaster construction and assembly flow diagram

Designers can develop block diagrams that show the flow for manufacturing a product. Such diagrams are useful in planning the manufacturing functions of purchasing (make-or-buy decisions), tooling, fabrication, assembly, inspection, and testing.

Designers can develop block diagrams that show the flow for manufacturing a product. Such diagrams are useful in planning the manufacturing functions of purchasing (make-or-buy decisions), tooling, fabrication, assembly, inspection, and testing.

Creativity is needed in developing solutions for the various parts of the manufacturing process. If large quantities of toasters are to be manufactured, then less expensive and more reliable assembly line and quality control techniques are investigated and trade-offs examined.

Manufacturers are willing to invest money in research and planning because the results of the studies and plans will provide worthwhile savings. The large savings obtained in the production of thousands of toasters can provide money for further studies and savings. Accountants amortize the cost of research and planning across all of the toasters produced. The money saved by designing more efficient and reliable production facilities lowers the production cost of each toaster. Thus, the selling price of each toaster may be decreased.

The toaster construction and assembly flow diagram (figure 9.7 on p. 227) becomes the production flow diagram. It identifies the individual steps required for efficient and reliable production. Such a diagram shows the sequence of assembly line relationships. Pugh (1993, p. 211) provides another example that involves a car door.

The production-line designers next commit to paper their ideas regarding the following:

PLANNING AHEAD ▶

- Who (workers or subcontractors) will produce which parts of the process?
- What is to be accomplished at each work station of the production line?
- When is the incoming material needed?
- Where will the incoming materials, partial assemblies, and final product be stored?
- Why should a new toaster design be considered?
- How can a new design meet consumer needs?
- How can a new design eliminate current production problems?
- How much could a new design contribute to overall company profits?

These questions—who, what, when, where, why, and how—are used to guide the investigators in a manner that will direct their work along productive paths.

The above questions may be used as a means of examining nearly any situation where several choices are available. The answers then provide the justification for investing money in the development and production of a new product.

SECTION 9.6   EXPANDING THE MODELING

This section continues the example of the use of block diagrams to help in effective design throughout the phases of a project. During the Conception Phase, the overall product or system block diagram is developed, as shown in the toaster example in section 9.5. The Conception Phase team may have devised a new approach to the solution. Sometimes, more than one block diagram is developed. Those diagrams are then expanded and evaluated during the Study Phase. From a block diagram, task lists, schedules, and cost data can be prepared.

Portions of the block diagram may be in question. Each questionable portion of the solution needs to be investigated before it is accepted as a workable solution. Whenever possible, the questionable portion of the solution should be either simulated and evaluated, or actually constructed and tested.

At the beginning of the Study Phase, information is gathered regarding existing models, which are used or modified for use whenever possible. For those portions of the block diagram where no existing models can be easily modified, new simulations or prototype models must be devised. The models are then used or tested and data are gathered. For models that were newly devised, data are often gathered over the life of the project during field trials of new units and during equipment and system operation. The information gathered can lead to modifications in later stages of the project and can be useful in preparing new designs of similar equipment.

The Design Phase takes advantage of the knowledge gained during the Study Phase modeling. The modeling results now point in the direction of the preferred design. If the design direction is still not clear, then the prototype of the entire system may need to be constructed. In this case, the remaining portions of the modeled system may then be constructed. They are connected to the partial-system prototypes previously constructed and evaluated. Additional data can now be gathered and examined for the entire system. Last-minute design changes may be needed prior to system construction.

Modeling is very important in the Study Phase. Models are prepared and results from tests of models guide the development of the design.

As computers become more sophisticated and can simulate real-world situations more accurately, the need for prototype models decreases. However, it would be incorrect to assume that all prototype models are unnecessary. First, most production tooling cannot be evaluated without the construction of a production prototype. Second, human-machine interaction that accurately reflects the effects of human boredom or tiredness upon the system cannot yet be simulated. Third, computer models are often not as accurate or all-encompassing as their designers may claim.

The Implementation Phase involves constructing and proving the actual system. Verification of the design for quantity production units may lead to some changes when the first production units reach the field. These changes are usually sent out to the users in "kit" form. More complicated changes to products may require field-service technicians to implement them.

There are times when it is necessary to run a simulation in real time. Such a simulation is connected to a portion of the prototype system. It must function at a speed that is at least as fast as the connected prototype. It must also be an accurate simulation; otherwise there will be GIGO (garbage in, garbage out). As an example, imagine a computer program for weather forecasting that operates so slowly that the actual weather has occurred by the time the weather predictions become available from the computer simulation.

Human beings have a wide variety of skills, abilities, and limitations. Machines with which humans interact must be designed with attention paid to these skills, abilities, and limitations.

Training can improve many of the skills and abilities of machine operators. Simulated emergencies can test and evaluate improvement in skills. Simulation of seldom-occurring problems can be helpful in further improving an operator's skills.

People react differently when rested, when tired, and when under stress. These three situations require study and, where possible, simulation, test, and evaluation.

According to Nadler (1970, p. 327), " . . . human agents represent any human resources used in or between any steps of the sequence or as part of the environment that aid in converting the inputs into . . . outputs without becoming part of the outputs." He identified the following typical human agents in a system:

Machine Operator	Tool and Die Maker	Fork Lift Truck Operator
Secretary	Computer Programmer	Janitor
Teacher	Nurse	Telephone Operator
Vice President	Supervisor	Engineer
Doctor	Clerk	Attorney

SECTION 9.7   MODEL INTERCONNECTING AND TESTING

When large systems are being modeled, initially subsystems may be devised separately. After each subsystem model is designed, constructed, debugged, and tested, then the subsystems must be interconnected—preferably two at a time.

Prior to designing the individual subsystem models, agreement must be reached on how they will eventually be interconnected. These interconnections are often referred to as **interfaces.** (Interfaces are the mechanical and electrical boundary for two or more pieces of equipment; cables, connectors, transmitting and receiving circuits, signal-line descriptions, and timing and control information are included.) Interfaces contain information related to electrical, mechanical, hydraulic, thermal, and environmental parameters. (The dynamic portion of these interfaces is known as **handshaking.**)

The individual subsystems are tested separately. Once the models are revised to reflect the final product as closely as possible, then it is time to interconnect the subsystems—two at a time. First the overall test information is gathered. Then the results are compared with the initially determined individual test results. When any deviation is noted, the models must be modified to reflect the corrections. The interfaces are now considered verified.

Now additional subsystems can be connected with the tested pair of subsystems. The procedure is repeated until all of the modeled subsystems are connected into the entire system, if this is feasible.

These subsystems are developed in a modular manner so they can be used later with other system designs with or without modification. Many people devote their professional careers to subsystem modeling and testing.

SECTION 9.8   MODELING THROUGHOUT THE PROJECT PHASES

A summary and sequence of the application of models is given below, by project phase. The flow is further summarized in figure 9.8.

Figure 9.8   Sequence of Modeling

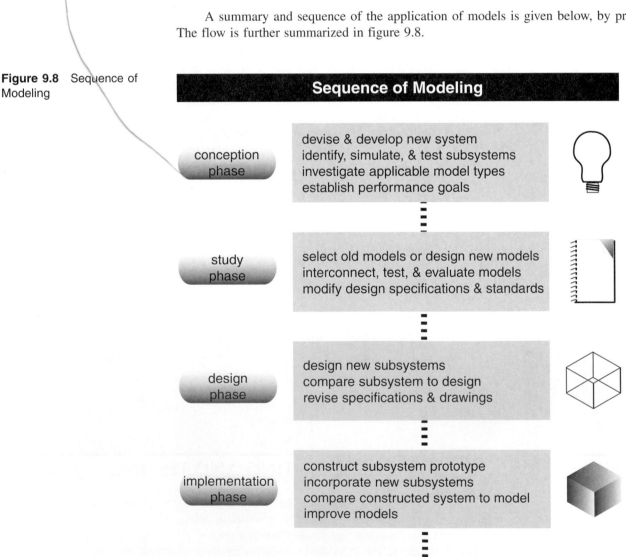

Sequence of Modeling

conception phase
devise & develop new system
identify, simulate, & test subsystems
investigate applicable model types
establish performance goals

study phase
select old models or design new models
interconnect, test, & evaluate models
modify design specifications & standards

design phase
design new subsystems
compare subsystem to design
revise specifications & drawings

implementation phase
construct subsystem prototype
incorporate new subsystems
compare constructed system to model
improve models

Conception Phase

1. Devise and develop the system block diagram.

   Use old blocks where possible; devise new blocks where needed.

   Select or devise new models for newly devised subsystem blocks.
2. Identify critical subsystem blocks that could cause system failure.

   Devise and expand critical-block solutions.

   > simulate and evaluate
   >
   > construct and test
3. Investigate model types that would apply to identified blocks.
4. Establish system and subsystem performance goals.

Study Phase

1. Select old models that apply to the identified blocks.

   Gather information from existing models.

   > simulate prototype
   >
   > construct prototype
2. Design new models that are needed.
3. Examine the effect upon the system of each model individually.
4. Interconnect models; test and evaluate model interactions.
5. Modify design specification based upon modeling results.
6. Establish system performance standards within that specification.

Design Phase

1. Design actual contents of each subsystem block.

   Construct and test new subsystem designs.

   Evaluate subsystem test results.

   Revise designs as necessary.
2. Compare subsystem design performance to design specification.
3. Revise Production Specification as necessary.
4. Revise Working and Detail Drawings as necessary.

Implementation Phase

1. Construct prototype of new subsystem designs.
2. Incorporate new subsystems into original system.
3. Compare total system performance with modeling results.
4. Improve the models so they are more realistic when used again.

As computers and the companion software become faster, more powerful, and more flexible, the application of simulations to modeling will increase. Therefore, it is advisable for persons interested in technical careers to learn one or more programming languages and become familiar with the capabilities of a wide variety of application programs.

Since it is not granted us to live long, let us transmit to posterity some memorial that we have at least lived.

Pliny the Younger (61 to 114 A.D.)

CHAPTER OBJECTIVES SUMMARY

Now that you have finished this chapter, you should be able to:

1. Define models and tell why they are used in development of designs.
2. List the kinds of models that are typically used in development of designs and give an example of each.
3. Explain why it is important to decide early in a project what to model.
4. Describe the criteria that a model should meet.
5. Define modules and tell why it is advantageous for models to be designed as modules.
6. Define block diagrams and flow diagrams and tell how they might be useful in preparing models for projects.
7. Describe what kinds of activities are appropriate in relation to models during the four phases of a project.
8. Define the study of human factors and tell why it is important in the design of projects.

EXERCISES

9.1   Describe the inputs, processes (functions), and outputs for a model assigned by your instructor.

9.2   Develop a modeling sequence that describes, in more detail, the applicable sequence by applying or adapting the blocks shown in figure 9.3 for a project assigned by your instructor.

9.3   Indicate which of the models indicated in figure 9.1 applies to your assignment for exercise 9.2 above.

9.4   Develop a flow diagram and/or block diagram for exercise 9.2 above.

9.5   Discuss how you would interconnect your model with those of other students involved in other parts of the same project. Note how the overall interconnection could be tested.

9.6   Write a paper describing how you could apply human factors considerations to your project.

9.7   Write a paper describing what you have learned from this text. Indicate how this text has improved your technical and personal skills, and how this text has affected your choice of a professional career, if applicable. Revise your resume to reflect your newly acquired skills and experience and, if necessary, your career-change direction.

APPENDIX A
Value Analysis and Engineering

HISTORY

Lawrence D. Miles developed value engineering in 1949 at General Electric. Materials were scarce because of the War effort. Thus, other materials were substituted to produce essential products. Informal analysis indicated that substitutes often performed better than the specified materials. From these findings, Miles organized and established the procedure, method, or approach known as **Value Analysis/Engineering (VA/E).**

The main focus of this method is to identify the functions of the product, design, or service under consideration. Alternate ways to fulfill each function are brainstormed. (See chapter 2.) Cost reduction is achieved through critical evaluation to find the best alternative (and least expensive way) for meeting those functions. Because the goal is to provide the essential functions at low cost while maintaining or increasing user acceptance and acceptable quality, the target of the Value Analysis effort is not fixed. In traditional cost reduction programs, the focus is merely on substituting less expensive parts into a product, without changing the basic configuration (Miles, 1972). Therefore, the objective of a Value Analysis study is to achieve the function with increased user acceptance and higher quality at a lower cost. This would be equivalent to increasing the ratio of function at a higher quality and greater user acceptability to the cost. An equation representing value can be expressed as

$$\text{Value} = \frac{\text{Function} + \text{Greater User Acceptance} + \text{Acceptable Quality}}{\text{Cost}}$$

In the 1960s, the Department of Defense embraced the Value Analysis concept and inserted incentive clauses into all government contracts. Programs meeting the guidelines and achieving savings on the contracts received extra money. Companies established new departments. Civilian projects in the private sector followed the military approach and established the process in the manufacturing and construction area (Dell'Isola, 1982).

As the military programs ended following the Korean and Vietnam conflicts, Value Analysis programs were scaled down or eliminated. However, as with all worthwhile techniques and ideas, the programs came back in all sectors of the economy, including manufacturing (both civilian and military businesses), service, financial, and construction. Reengineering and Concurrent Engineering concepts and programs brought back Value Analysis. The process starts at the beginning and continues throughout the design cycle. This replaced a design review after the product, process, or system was nearing completion or in production.

The term *Value Analysis* will be used in the rest of this appendix to describe the *process*. It employs teams, with three to seven people on each team. Mixed groups are composed of people from different departments. Individuals familiar with the problems and others with no previous background appear to be the most successful team members. Initial communications barriers may exist. An experienced leader is essential. The total procedure, with examples, is included in the following paragraphs.

THE VALUE ANALYSIS PROCESS

Poor value is designed into a product when functions are not analyzed. This occurs because individuals follow tradition. They lack information or make the wrong assumptions or do not think creatively. Value Analysis does not accept a designed part, process, or system; it stresses cost reduction by defining the **function.** A product or service increases in value when its basic functions, as well as its secondary functions, perform better than those of competing products or services. Value analysis answers the following questions (Brown, 1992):

- What are the parts, processes, or pieces?
- What does it do?
- What does it cost?
- What else will do the job?
- What will that cost?

Value Analysis has been successfully used in all engineering disciplines and specialties, and in organizations within the human and social services fields. The process is a structured sequential plan or strategy that consists of four phases: the information phase, the speculation phase, the evaluation and analysis phase, and the implementation phase (Miles, 1972; Fasal, 1972; and Mudge, 1971). It is best taught and learned by limited lecturing and with many exercises, demonstrations, and projects. Each of these phases will be discussed in more detail.

THE INFORMATION PHASE

In creative problem solving, the emphasis is on the creative thinking activities: idea generation and evaluation of creative ideas. In Value Analysis, as in any design process, the first (and the hardest) step is problem formulation or definition. The four tasks of the information phase are as follows:

1. Data collection
2. Function identification
3. Cost analysis
4. User reaction surveys

Task 1. *Data collection* focuses on identifying accurate user data about the product, part, object, or service. These words will be used in different situations but will apply to all problems. Market surveys on consumer satisfaction and attitudes, user expectations and experiences, customer wants and needs, as well as competitive comparisons are very important in obtaining complete and accurate information. Three distinct groups of users must be consulted:

1. Consumers, customers, or clients who have experience with the product
2. Managers or administrators who are involved in supervising the production or acquisition of the product
3. People who are directly involved in the manufacture or execution of the product.

Task 2. *Function identification* is a difficult and time-consuming task. The primary rule is:

ALL FUNCTIONS ARE EXPRESSED IN TWO-WORD (VERB-NOUN) PHRASES.

Functional analysis is similar to the development of task descriptions for projects such as those described in chapter 7. All functions that the product does or is expected to do are expressed in verb-noun phrases in Value Analysis studies. Verbs are action words (see appendix B). Nouns are measurable and must be used in primary functions. Nonmeasurable nouns represent basic or supporting functions.

First, the primary function must be identified. This function is the reason for all other basic and supporting functions. For example, the primary function of a pencil is to make marks; the primary function of a lawn mower is to cut grass; the primary function of an engine is to supply power; the primary function of a window is to provide light.

Next, basic and supporting functions must be identified. This activity is surprisingly difficult and requires a lot of brainstorming by the teams. Basic functions are critical to the performance of the primary function. Supporting functions are subordinate to the basic functions. In many products, these are "perks" added to increase buyer appeal. However, they can add significantly to the cost of the product. Basic and supportive functions are then subdivided into secondary and tertiary (third level) groupings. In brainstorming, combining ideas is encouraged. In function identification, separation is the goal.

A chart is prepared that shows how the different functions are connected. It is known as the **FAST (Function Analysis System Technique) diagram** and is organized horizontally. Each phrase branches to answer *how* the function directly adjacent to it is to be performed; each phrase along the chart answers *why*. For example, in figure A.1 "scribe arc" is the primary function of the compass. To perform that function, the secondary functions of "position" and "rotate the marker" must be accomplished. To position marker, we must "have marker," "hold marker," "separate legs," and "anchor center" (tertiary functions).

This chart can be developed as a group activity. Each phrase is written on a small card. These cards are moved around and new phrases are added. The group must be satisfied that all the functions of the product have been identified and organized in their correct relationship on the chart.

For example, the primary function of a pencil is to make marks. Several secondary functions are identified. One of these is to contain advertising and another is to correct mistakes; an eraser is included. Tertiary functions are not necessary in a design as basic as a pencil. A completed FAST diagram—figure A.1—shows a sketch of a compass, all its functions, and the boundary conditions (scope) of this study (Salvendy, 1992).

Let's look at another example. The primary function of a window is to provide daylight. Secondary functions are to allow air circulation, provide security, and keep out undesired weather. A tertiary function would be to repel insects.

Task 3. *Cost analysis* is very detailed and analytical, and it can be tedious and boring. Various strategies and costing procedures are available for this task. The most appropriate or easiest one should be used to yield a reasonable allocation of cost to each item in the function tree. Labor, material, design, and overhead costs must be included. Other factors may be involved as well. Accounting specialists use a variety of equations to allocate costs. Accounting functions within the organizations should be consulted. The use and allocation of the data, however, must be relevant to the function. All costs must be tabulated and included under the appropriate function.

Task 4. *User reaction surveys* are a diagnostic tool. When the function tree has been completed, the team must determine which of the functions are most important to product performance and customer or client acceptance. Feedback should be obtained from all stakeholder groups. Functions that are relatively high in cost but are judged unimportant by

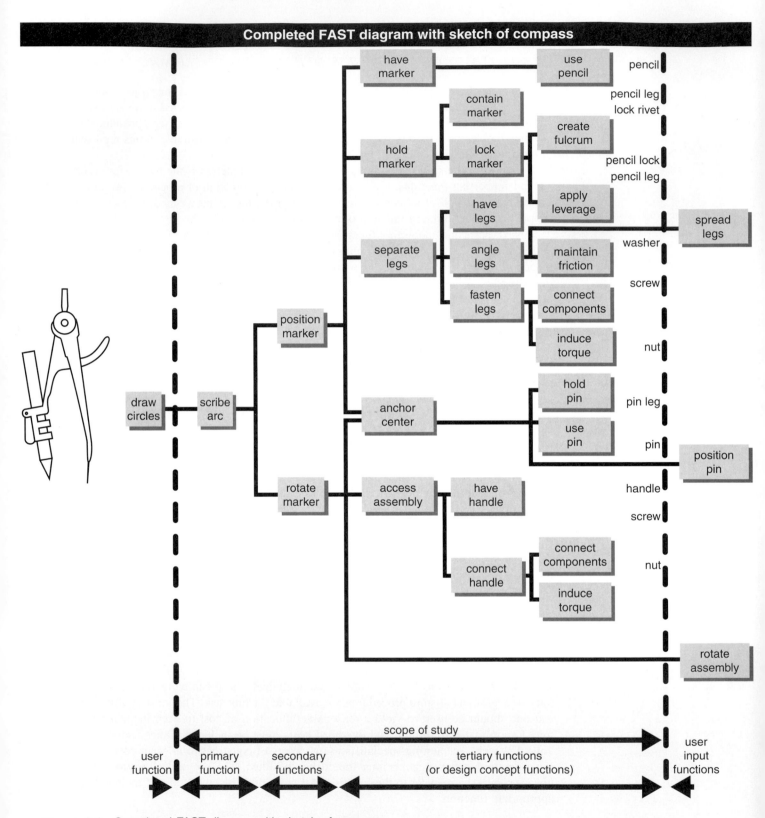

Figure A.1   Completed FAST diagram with sketch of compass

Salvendy, G., *Handbook of Industrial Engineering,* © 1992. Reprinted by permission of John Wiley & Sons, Inc.

stakeholders are prime candidates for cost reduction. This identification of high, low, and indifferent acceptance (or A, B, C classification) of function categories is essential to the next phase. Development of user-desired product criteria is an activity that requires creative thinking and data analysis.

SPECULATION OR SEARCH PHASE

Brainstorming is used to generate ideas. The questions to be answered about each function are:

- What else could do the job?
- In what other ways can each function be performed?

Great emphasis is placed on deferred judgment; the group members are not allowed to assess, evaluate, or criticize. Individuals are encouraged to note their ideas before verbal brainstorming begins. All ideas are recorded. These ideas are shared, additional ideas are generated, and all ideas are written on large charts for all to see. (Refer to chapter 2 for additional guidance on brainstorming.)

EVALUATION AND ANALYSIS PHASE

Brainstormed ideas are evaluated in terms of cost, feasibility, and other relevant criteria. The key objective for this phase is to improve ideas. Ideas are only eliminated if a better solution is developed. Although critical judgment is used, people are encouraged to maintain a positive attitude. The cost figures and the importance ratings developed earlier are very useful for focusing this effort. This phase is very important for those functions that have a high acceptance or a high cost.

IMPLEMENTATION PHASE

The objective of this phase is to sell the recommended improvements to stakeholders for the product resulting from the Value Analysis study. A proposal is prepared that demonstrates that the proposed changes are technically feasible and meet the management objectives. Substantial cost savings are also documented. Schedules and budgets for implementation need to be prepared, as well as an assessment of the new skills and employee training that will be required. Long-range and short-range feedback and monitoring plans must be established to evaluate the results of the effort. The proposal presentation is then made to management and administrators using good presentation techniques and visual aids. The "sales" presentation should be brief and should focus on the highlights.

Implementation of creative problem solving through value analysis requires support from top management. Cooperation and teamwork among departments make it a multidisciplinary approach. Value Analysis requires the allocation of necessary resources. A variety of forms, data collection and analysis procedures, and computer programs can be developed so the process will be more efficient (Brown, 1992).

ADDITIONAL THOUGHTS AND IDEAS

Some supporters of Value Analysis recommend that the Implementation Phase be split into three phases:

1. Program Planning involves summarizing additional sources of ideas or solutions from plant personnel and vendors.
2. Execution involves preparing and evaluating additional proposals.
3. Summary and Conclusion is when the formal report and presentations are given.

Training employees in creative thinking and problem solving is effective and productive only when top management is committed to these activities and provides a supportive environment. Special techniques are available to make you become a better problem solver without having to develop a new procedure for your work. See Allen (1962), Lumsdaine, E. & M. (1990), Pugh (1993), and von Oech (1993).

SUMMARY

The following techniques or phrases are used in any Value Analysis study. Use them in your studies as a checklist.

- Set up the Value Analysis job plan.
- Practice creative thinking to overcome roadblocks; then refine the new functions.
- Evaluate each function by comparison.
- Get data and facts from the best sources using actual costs and standards.
- Put dollar signs on the main idea and key tolerances.
- Use specialty products, materials, and processes.
- Spend company money as you would your own.
- Work on specifics, not generalities.
- Use your own judgment with good human relations.

The Value Analysis process is simple and organized. Implementing it will lead to an improved design at a lower cost.

APPENDIX B
Objectives, Task Descriptions, and Active Verbs

WRITING OBJECTIVES FOR TASKS

Written objectives are used to guide a project. Some describe activities that have been accomplished many times previously. If this is the case, then the previously written and used objectives may be applied directly or modified to suit the current project. Also, written objectives may describe a new, unique, and untried activity to be undertaken. It may be uncertain that the goals of the activity can be met without a further, more precise description.

The written-objective technique was devised under the direction of the late Admiral Rickover. He was assigned responsibility for the design, development, construction, and testing of nuclear-powered submarines. At approximately the same time, Robert Mager developed the written-objective technique for education.

Many school systems use the written-objective technique to define the material to be taught and learned. At the end of the school term or year, special exams are given to determine both the new abilities and specific shortcomings of individual students. These exams also evaluate the instructor's success in teaching the material.

In business and industry, written objectives are prepared to describe the intended outcome, or result, of the work to be performed. They describe what the results should look like, or act like, when the tasks have been completed successfully. They are included either within a Specification document or in a separate Project Task Description document.

The work to be performed is divided into tasks that are included in the Project Task Description. Today, objectives are a key portion of each task in a project. Many objectives are written with the measurable goals to be achieved clearly stated. Results are judged by how closely the design agrees with the numbers that are a part of a measurable objective. The use of numbers permits persons contracting for the construction of equipment and systems to evaluate the progress of the engineering and management team performing the design. There are, therefore, fewer lawsuits when objectives are defined and negotiated in advance.

An objective describes the destination, or end result, of a task. Task analysis documents describe the details, such as: when the task is to begin (the origin), the work to be performed (the route), and the end result (the destination). An objective indicates the direction to be pursued without restricting the designer to a specific final result. An objective may also mention related goals. Thus, all persons involved can verify that the documented goals have been achieved.

The Mager Six-Pack (1996) is an excellent guide to writing objectives. It contains both instructions and examples, with built-in feedback.

Note that a written objective contains the following descriptions:

* The conditions under which the task will be performed
* What you want someone to be able to do (the necessary skills)
* The criteria that describe successful performance; the definition of the *measurable* final result(s)

The objective within a task should allow evaluation of the end result. All persons involved will know when the documented goals have been achieved because the objective is written in a clear manner.

Figure B.1 Three types of objectives

Three Types of Objectives

behavioral objectives
qualitative goals
active verbs

performance objectives
quantitative goals
active verbs

technical objectives
quantitative goals
active verbs

There are three types of written objectives (figure B.1):

1. Behavioral objectives broadly describe the goals of a project or portion of a project and cannot be measured with numbers.
2. Performance objectives describe the goals of a project or portion of a project by means of numbers. All persons involved agree that the goals can be measured with numbers to determine that they have been satisfactorily achieved.
3. Technical objectives precisely describe, in technical terms, the work to be accomplished. They can be measured with numbers.

An objective that describes the desired end result is a **terminal** objective.

Behavioral objectives provide the qualitative goals for a project. An example of a behavioral objective for the activity center would be the following:

Design an activity center for children that provides a pleasant atmosphere leading to friendly interactions with their companions and their adult supervisors.

Performance objectives should be key portions of a task description. Equipment and procedures are judged by how closely their design agrees with the numbers that are a part of each performance objective. They permit persons contracting for equipment and systems to evaluate the progress of the engineering and management team performing the design. Therefore, there should be fewer recriminations, hard feelings, and lawsuits when objectives are defined in advance: Everyone agrees in advance on what the outcomes should be. For the activity center, an example of a performance objective would be the following:

There shall be a minimum of 1' clearance between the exterior walls of the foundation and the interior walls of the excavation.

The specifications for projects contain many technical objectives. (See figure B.2.) Terms are used that are meaningful to those professional persons involved in achieving those objectives. Designers have flexibility of choice within the broad constraints provided by the technical design criteria.

For the activity center, here is an example of the evolution of a technical objective through the various phases of a project. For the Conception Phase, the wording is often most general:

Select flooring whose cost is less than 50¢ per square foot.

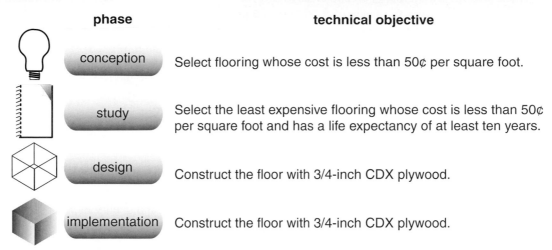

Figure B.2   Development in the content of objectives through the project phases

For the Study Phase, the technical objective should not include a conclusion or result.

Wrong:
Construct the floor with $^3/_4$-inch CDX plywood.

Right:
Select the least expensive flooring whose cost is less than 50¢ per square foot and has a life expectancy of at least 10 years.

With the more general objective, during the Study Phase, designers should evaluate only those types of flooring that will be within the price and life-expectancy limits given in the Design Specification being developed.

The "wrong" Study Phase technical objective provides both the direction and the result or conclusion. The "correct" Study Phase technical objective indicates only the direction toward a conclusion.

The Study Phase "wrong" objective now represents the correct wording for the Design and Implementation Phases:

Construct the floor with $^3/_4$-inch CDX plywood.

An objective is written as an instruction. An objective starts with an active verb and is a complete (standalone) sentence. Words that are used in an objective should be understandable to all persons assigned to study, design, or implement the design product, process, or procedure. A list of active verbs appears at the end of this appendix. These active verbs are used in the writing of behavioral, performance and technical objectives.

WRITING TASK DESCRIPTIONS

A task description focuses upon *what* is to be accomplished. As an example, the activity center foundation is to be constructed according to the foundation portion of the Production Specification. The Project Task Description describes in writing how the foundation work is to be accomplished (appendix C gives the final specification). A well-written task description leaves no question regarding what work is to be performed and how that work is to be performed.

Why have task descriptions? Because there are more wrong ways to do something than right ways. Verbal instructions are sometimes given for work to be performed. However, such verbal instructions may be vague, which in turn may lead to misunderstandings and incorrect performance of work. Once a project is initiated, someone must decide what work is to be done, who is to do it, when it is complete, and whether it has been performed satisfactorily.

A task is a small part of a project. Therefore, a task will have many of the attributes of a project. It may be considered a miniature project. The work to be performed must be defined in sufficient detail to allow the workers to complete the task with a minimum need for further instructions.

A task description will contain the following:

- The name and objective of the task
- The relation of the task to the overall project
- A description of the work to be done
- The specifications relating to the materials and quality of work
- A schedule for the performance of the task

Two examples of the *technical* content for a task description for the activity center are given below.

Prepare Site
The site shall be surveyed, by Registered Land Surveyors or Civil Engineers, to establish grade levels of the basement floor. A stake shall be placed at each corner of the foundation with "depth of cut" noted on each stake. Batter boards shall be installed for each corner and marked with the location for each foundation wall. See figure B.3.

Excavate Site
The excavation of the basement and foundation footings is to be performed next. There shall be a minimum of 1' clearance between the exterior walls of the foundation and the interior walls of the excavation. If ledge outcroppings are discovered, then they are to be drilled and blasted to remove the ledge.

The above paragraphs give a **technical** description of the work to be accomplished. Next, let's expand the latter paragraph to include all the information required of a total task description.

Figure B.3 Foundation wall positioning technique

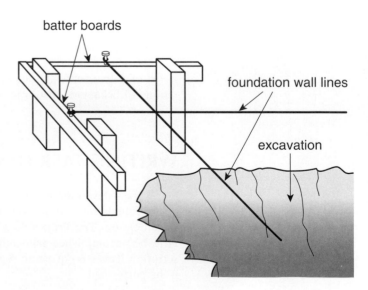

Excavate Site

   —This is the name and objective of the task.

Site excavation shall follow completion of site preparation and precede the erection and capping of the foundation.

   —This describes the relation of the task to the overall project.

The excavation of the basement and foundation footings is to be performed next. There shall be a minimum of 1′ clearance between the exterior walls of the foundation and the interior walls of the excavation. If ledge outcroppings are discovered, then they are to be drilled and blasted to remove the ledge. (See the specification in Appendix C, Cast-in-Place Concrete, paragraph G.)

   —This is a description of the work to be done, and it includes a reference to the paragraph in the specification relating to the materials and quality of work.

Site excavation shall be initiated during the first week of the project and shall be completed prior to the end of the second week of the project.

   —This is a schedule for the performance of the task.

Note that the first task described above (Prepare Site) directly leads to the second task (Excavate Site). All other tasks connect to at least one previous task and lead to at least one other task. Once the last task is completed, the inspection would be performed according to the final criteria for the project. (Other inspections may be required during project performance.)

When people are assigned to work on a project, each person is given oral and written instructions describing the work to be performed. All participants should be given enough information so that they understand how their work contributes to the completion of the overall project. Then they will realize the importance of their contribution to the total project and will be motivated to do their best work. Verbal instructions are often used for this purpose. However, it is better to also provide written instructions so that each worker may refer to them as needed to be certain that the work is being completed in accordance with the contract.

The Project Task Description should be brief and concise. It describes work that is extensive and complicated enough to require the coordinated efforts of several people. Work to be performed, as described within the task description, must be measurable.

Many engineers and technologists write poorly. Employers complain bitterly that their technical employees cannot communicate effectively and in writing, considering the salaries they are paid. This communication problem includes both their technical supervisors and their higher-level nontechnical executives. The U.S. Accrediting Board for Engineering and Technology (ABET) has indicated that correcting this communication deficiency is one of its top priorities.

A LIST OF ACTIVE VERBS

Objectives start with active verbs and are a complete (stand-alone) sentence. Words that are used in an objective should be understandable to all persons assigned to study, design, or implement the design.

Below is a list of active verbs that might be used in task descriptions. It is considered neither perfect nor complete. Add to this list any new active verbs that you "discover."

absorb	accept	access	accommodate
accompany	accomplish	achieve	acquaint
acquire	activate	adapt	add
adhere	adjoin	adjust	advance
advise	advocate	affect	agree

align	allot	allow	alter
amortize	analyze	announce	appear
apply	appraise	appropriate	approve
approximate	arrive	ascertain	assemble
assign	assist	associate	assume
assure	attain	authorize	avail
avoid	award		
backfill	balance	base	become
begin	believe	bend	bid
bind	blast	block	boil
boot	bore	bounce	brace
break	budget	build	
calculate	catch	cause	certify
change	charge	chart	choose
collect	combine	commit	communicate
compare	compensate	complain	complete
comply	comprehend	compute	conceive
concentrate	concern	condense	conduct
confer	configure	confirm	conflict
connect	consider	constrain	construct
contact	contain	continue	contract
contribute	control	convert	convince
coordinate	correct	correlate	cost
create	cut		
debate	decide	decrease	define
delay	delegate	deliver	demonstrate
deny	depend	derive	describe
design	designate	desire	determine
develop	devise	differentiate	direct
discover	discuss	display	disseminate
distinguish	distribute	disturb	divide
do	document	draft	drill
earn	edit	effect	elect
elevate	elicit	eliminate	employ
empty	emulate	enable	enclose
encourage	engrave	enhance	enjoy
enlarge	enlist	enroll	ensure
enter	equip	erase	erect
escape	establish	estimate	evaluate
evoke	evolve	examine	excavate
exceed	excel	exchange	excite
exempt	exercise	exert	exhibit
expand	expect	expel	expend
experiment	explain	explore	expose
extend	extract		
fabricate	feel	finish	fit
focus	forecast	form	formulate
forward	frustrate	fulfill	function
gain	gather	generate	give
graph	grow	guess	guide
hire			
identify	ignore	illustrate	implement
improve	include	incorporate	increase

indicate	inform	initiate	insert
inspect	install	instruct	integrate
interact	interconnect	interest	interpret
intertwine	invest	investigate	invite
involve	irritate	issue	
jack	jam	join	judge
justify			
keep	kick	kill	know
label	lace	lay	lead
learn	leave	lend	lengthen
let	level	license	lift
light	limit	list	loan
locate	lock	lower	
mail	maintain	make	manage
manipulate	manufacture	map	match
mate	measure	meet	memorize
mend	merge	minimize	miss
mix	model	modify	monitor
mop	motivate	mount	move
mow	multiplex	multiply	
need	neglect	negotiate	note
notify	number		
observe	obtain	occur	offend
offer	operate	order	organize
originate	outline	overlap	oversee
paint	participate	pay	perceive
perform	permit	place	plan
precede	predict	prefer	prepare
present	preserve	prevent	proceed
process	produce	progress	project
promote	propose	protect	prove
provide	purchase		
react	read	realize	recall
receive	recognize	recommend	recompute
reconstruct	record	reduce	refer
reference	refine	reflect	regard
register	reinforce	relate	remain
remember	remove	repair	repeat
replace	report	represent	reproduce
require	reschedule	research	reserve
resolve	respond	retain	review
revise	reward		
satisfy	save	scan	schedule
search	select	sell	separate
share	shorten	sign	simplify
simulate	sketch	solder	solve
specify	stabilize	standardize	start
state	stop	store	structure
study	subdivide	submit	substitute
subtract	summarize	supervise	supply
support	synthesize		

take	telephone	terminate	test
trace	train	transfer	transform
understand	unite	update	use
utilize			
vanish	vary	verify	visualize
want	waste	watch	waterproof
win	wish	withdraw	withhold
work	write		

APPENDIX C

Specification for the Town of Bedford, Massachusetts: "An Activity Center"

GENERAL SPECIFICATIONS

For the purposes of the contract and this specification, the following terms shall have the meanings set forth below.

Both organizations and individuals are identified since both are considered responsible should legal action be required.

A. The term "client" refers to Town of Bedford, in Massachusetts, whose principal Town Office address is 44 Mudge Way, Bedford, MA 01730. The authorized representative for the Town of Bedford is James T. Williams, Superintendent of Public Works.

B. The term "contractor" refers to the corporation, identified as such in the contract, responsible for execution of the work contracted for by the client. The contractor is Northeastern University, whose principal address is 360 Huntington Avenue, Room 467, Snell Engineering, Boston, MA 02115-5096. The authorized representative for Northeastern University is Thomas E. Hulbert, Professor of Industrial Engineering.

The contractor is responsible for all subcontractors.

C. The term "subcontractor" means, without limitation, any firm or person working directly or indirectly for this contractor, whether or not pursuant to a formal subcontract, that furnishes or performs a portion of the work, labor, or material, according to the drawings and specifications.

The contract will refer to several documents, including this specification. These documents interact with one another. Therefore, for legal reasons, their interrelation needs to be described.

D. The term "contract" means the construction agreement signed by the client and the contractor, which includes this specification and all other documents listed as contract documents.

E. The term "work" includes all labor necessary to produce the construction required by the contract, all materials and equipment incorporated or to be incorporated in such construction, and other items necessary to produce the construction required by the contract.

GENERAL REQUIREMENTS

The contractor and subcontractor must carefully study all applicable documents.

A. The contractor warrants and represents that it has carefully examined all of the plans and specifications, and has satisfied itself as to the conditions and difficulties likely to be encountered in the construction of the work. The contractor agrees that any conditions or difficulties that may be encountered in the execution of the work, which should have reasonably been discovered upon such examination, will not constitute a cause for modification in the contract amount, or a termination hereof.

Public agency rules and regulations indicate which permits are the responsibility of the client and which rules and regulations are the responsibility of the contractor.

B.   The contractor shall obtain and pay for all permits (except for building and subsurface wastewater disposal system permits), licenses, certifications, tap charges, construction easements, inspections, and other approvals required, both temporary and permanent, to commence and complete the work. No additional cost to the client shall be made because of this requirement.

C.   In the event of nonacceptance of the work by any regulating authority, the contractor shall perform the additional work required to bring the work into a condition of acceptance, and such additional work shall be performed by the contractor without any further cost to the client.

D.   The contractor shall protect the work and all nearby property from loss or damage resulting from its operations, and in the event of such loss or damage, shall make such replacements or repairs as required by the client without additional cost to the client.

The withholding of money is sometimes used by the client or other funding agencies to ensure the completion of work on time and according to the applicable documents.

E.   The contractor shall furnish the client any permits and warranties required or obtained before final payment shall be due to the contractor. All work shall comply with applicable state and local codes and ordinances.

F.   All work shall be performed in a good workmanlike manner, using accepted engineering practices and accepted construction methods.

G.   All materials and products are to be installed with strict adherence to manufacturers' instructions in order to validate warranties.

Contractors must have some flexibility because some materials and equipment may not be available at the time that actual purchase is attempted.

H.   The contractor shall have the option to use material or equipment equal to or better than materials specified in these documents. Samples shall be submitted to the client for approval before any substitutions are made.

INTENT OF CONTRACT DOCUMENTS

As stated previously, the contractor and subcontractors carefully study all applicable documents because they are held responsible for their content—even if errors exist in these documents.

A.   The agreement and each of the contract documents are complementary, and they shall be interpreted so that what is called for on one shall be binding as if called for by all. Should the contractor observe any conflicts within the contract documents, it shall bring them promptly to the client's attention for decision and revision.

B.   The contract documents will describe the scope of work for the project. A duplication of work is not intended by the contract documents. Duplications specified shall not become a basis for extra cost to the client.

SUBCONTRACTORS

The client does not want to be involved in a disagreement between the (prime) contractor and its subcontractors. Management of subcontractors is, in this contract, the contractor's responsibility.

A.   The contractor, at its discretion, may elect to have certain portions of the work required to be performed by others under subcontract.

B.   No provisions of the contract, nor of any contract between the contractor and subcontractors, shall be construed as an agreement between the client and the subcontractors.

C.   The contractor shall be fully responsible to the client for the acts and omissions of the subcontractor, as it is for the acts and omissions of its own employees.

PROJECT MEETINGS

Interaction among individuals and groups is vital in order to achieve technical understanding. English is not a precise language. Thus, to avoid misunderstandings, it is necessary to communicate verbally as well as via paper documents. The importance of such meetings includes identifying problems and agreeing upon the action that must be taken. See chapter 8.

A.   Periodic project meetings during construction will be scheduled and administered by the contractor with the client to review work progress, note field observations and problems, and coordinate projected progress. Written minutes of each meeting shall be prepared by the client. They shall be distributed to all attendees within five (5) working days of the meeting, documenting all decisions and action items agreed upon during that meeting.

CLEANING

A.   The contractor shall at all times keep the premises free from accumulation of waste materials or rubbish caused by its employees, contractors, or their work. At the completion of the work, the contractor shall remove from the building and site all rubbish (including damaged trees and any other stumps), tools, and surplus materials, and shall leave the workplace "broom clean" unless otherwise specified.

WORKERS' COMPENSATION AND INSURANCE

A.   The contractor shall, at its own expense, comply with all the provisions of the Workers' Compensation laws of the Commonwealth of Massachusetts and shall supply the client with documentation verifying same prior to commencing work.

B.   Before commencing work, the contractor shall deliver to the client an insurance policy whose form and company are satisfactory to the client.

PROJECT CLOSEOUT

Formal documents are often necessary to "prove" that the contractor is the responsible party and the client is only responsible for receiving (accepting) the results and paying for the associated services.

A.   The contractor will be required to deliver to the client affidavits of payment of debts, release of liens for final payment, and all maintenance and operating manuals, warranties, and guarantees, if any are available or required.

GENERAL SITE WORK

Now the more technical wording begins!

A.   The contractor shall excavate as indicated on the drawings. Earth banks shall be braced against caving in the work area. The bottoms of all footing excavations shall be level, and on solid undisturbed earth or rock ledge. Excavations are to be kept free of standing water.

B.   All obstructions to the progress of work area are to be removed. Obstructions include: two trees next to the road and a significant portion of a lilac bush. These obstructions will be marked by the client with white cloth and orange spray paint.

C.   Property damaged during construction shall be repaired by the contractor. Any damage to adjacent property is to be repaired at the expense of the contractor.

D.   All disturbed topsoil in the building area and for a distance of 10′ beyond the outside walls of the building is to be stockpiled on the lot.

E.   Demolition and removal of any existing dwelling shall be included in the contract price.

EARTHWORK

A.   All excavating, filling, backfilling, and grading required to bring the entire project to the finish grades shown on the drawings shall be performed using the stockpiled topsoil and other material.

B.   Fill shall be free from debris, large stones (over 6″ in diameter) and perishable, toxic, or combustible materials.

C.   A minimum of 6″ of crushed stones or graded gravel shall be provided as an underlayment under all exterior slabs on grade, compacted to 90%.

D.   All rocks and boulders greater than 1 cubic yard shall be buried a minimum distance of 8′ from the structure or removed from the site. If there is not room to bury the rocks and boulders nearby, then they must be removed to another location. A state environmental permit may be required. Responsibility for these costs should be determined prior to contract initiation.

E.   Backfill around exterior foundation walls shall consist of clean and well-compacted material with stones that do not exceed 4″ in any dimension.

F.   Remainder of site shall be graded to ensure drainage of surface water away from building.

G.   Any excess soil is to be removed from the site upon completion of work.

This is a qualitative statement. The radius of the slope curve will have to be determined when the grading is in progress.

H.   Grades not otherwise indicated on the plans shall be uniform levels or slopes between points. Changes in slopes shall be well rounded.

LANDSCAPING

A.   Topsoil shall be provided and spread to a minimum depth of 4″ in areas to be seeded.

B.   All areas not paved, planted, or occupied by buildings shall be seeded with grass. Seeding, fertilizing, and liming shall be carried out according to standard practices. Seed shall be supplied by the client. The seeded area shall contain a minimum of 4″ of topsoil.

SITE UTILITIES

A.   The contractor shall furnish all materials, labor, tools, and related items to complete the construction in every detail and leave in working order all utilities as called for herein and/or shown on the construction documents. This includes, but is not limited to, sewer connection, water service, natural gas, electric service, and telephone.

CONCRETE FORMWORK

A.   The contractor shall provide all labor, material, and equipment required to complete all the concrete work. All walls shall be plumb and level.

B.   Formwork for concrete shall conform to the shape, lines, and dimensions as specified for the foundations, walls, and footings, and shall be sufficiently tight to prevent mortar leakage.

C.   All forms shall be adequately braced to maintain their positions during pouring operations.

CAST-IN-PLACE CONCRETE

A.   Foundation walls and concrete footings are to be of sizes shown below:

1.   Foundation footings—18″ × 10″ placed and poured at one time

2.   Lally column footings—36″ × 36″ × 12″ to carry the load of the center partition

3.   Chimney footing—pour simultaneously with foundation footing

4.   Foundation walls—8″ thick with two rows of rebars, with a clear basement height of 8′0″ from top of basement floor to bottom of floor joist

5.   Insulate interior of concrete basement walls using 2″ R-10 styrofoam

B.   The basement slab on grade shall be 5″ thick, nonreinforced, and troweled to a hard, smooth finish. The basement floor shall be installed prior to commencement of wall framing—depending upon the external air temperature. The slab grade is to be set to ensure a level floor that does not deviate more than $\frac{1}{4}$″ in 10′. A 4″ floor drain shall be installed as shown on the plan.

C.   Footing concrete shall be Type I and have a minimum 28-day compressive strength of 2500 psi.

D.   Foundation and floor slab concrete shall be Type I and have a minimum of 28-day compressive strength of 3000 psi.

E.   The foundation footing shall be installed below the depth of frost penetration, and shall be coated with a mastic sealant. Insulation shall also be installed on the interior of the foundation walls, having a minimum "R" factor of ″8,″ and applied with a mastic or nails to hold it in place.

There are times when the reasons for technical information, such as the 4″ slump, are explained. This usually occurs when the contractor may want flexibility and needs to know why the 4″ slump was specified.

F.   Footing and foundation concrete shall be poured with a 4″ slump. This is a fairly stiff mixture and will require a vibrator. The vibrator, if not used properly, would defeat the use of a 4″ slump and cause segregation. An alternate solution would be to add plasticizer to concrete that would increase the slump to a range from 3″ to 8″ and not reduce the strength of concrete. The purpose of a 4″ slump is to obtain a uniform mixture of concrete and create a tight wall that will resist leakage.

If the cost of ledge removal is not acceptable to the client, then an alternative approach (such as omission of the basement) may be necessary. This is normally decided "in the field," which means at the construction site.

G.   If ledge is encountered during footing preparation, the foundation and/or footings may be anchored to solid ledge. Ledge removal is not part of this contract. The cost of ledge removal shall be borne by the client, with that cost determined prior to performance of work.

H.   Sill plates are to be secured to the foundation with $\frac{1}{2}$″ diameter, 12″ long anchor bolts, embedded no less than 8″ into the concrete and secured with washers and nuts.

DAMPPROOFING

A.   Provide a 6-mil polyethylene vapor barrier under the floor slab, extending the barrier to cover the top of the interior footing.

B.   Break all form ties; seal with hydraulic cement.

C.   To prevent capillary action of the concrete foundation with the soil, apply a generous coating of asphalt (bituminous) emulsion to the outside of the foundation below grade. The emulsion may be applied hot or cold, and brushed, sprayed, or troweled onto the foundation.

Trade names for materials are often selected because of the material's reputation for long-term, high-quality performance.

D.   Place polyethylene sheeting over the emulsion for the entire foundation. Trade names of acceptable cross-laminated polyethylene wraps are Tu-Tuff, Super Sampson, Griffolyn, and Cross-Tuff. Extend the emulsion to the top and side of the exterior footing, and extend the polyethylene to cover the top and side of the footing.

SUBSURFACE DRAINAGE SYSTEM

A.   Around the exterior footing, install 4″ perforated plastic pipe embedded in $\frac{3}{4}$″ or 1″ stone. The pipe must be 4″ below the bottom of the basement slab and pitched downward, toward the outside wall, at a minimum slope of at least 1″ in 10′. Place 6″ of stone under and around the pipe; cover the pipe with 3″ of pea stone. Cover the pea stone with 6″ of straw or hay, firmly packed, or use an external synthetic fabric drainage envelope filter sleeve. (This sleeve is produced by American Drainage System and is known as ADS Drain Guard or ADS Sock. It comes with a drain envelope/filter installed and requires only a few inches of stone on which to lay the pipe.) Extend and pitch the pipe so the water will drain away from the basement for a distance of at least 50′.

B.   In the interior of the basement area, install 4″ of perforated plastic pipe around the perimeter of the foundation footing. Extend the pipe through the footing to connect to the pipe installed on the exterior side of the footing. Fill the basement subfloor with a minimum of 6″ of $\frac{3}{4}$″ or 1″ process stone.

FRAMING AND CARPENTRY

A.   The sills should not be installed unless the foundation does not deviate more than $\frac{1}{2}$″ on any plane. Install the sill seals underneath the 2″ × 6″ pressure-treated sill. The sills are to be pressure treated with chromated copper arsenate (CCA) and are to have a 15-year warranty life.

Correction of nonvertical or nonhorizontal foundations is very expensive and complicated. Therefore, contractors are very careful when installing the foundation forms. Consult a specialized text on this matter.

B.   All structural members shall be kiln-dried construction grade spruce or fir. The following dimensioned lumber shall be used as specified:

1.   Floor joists—2″ × 10″

2.   Ceiling joists—2″ × 8″

3.   Roof rafters—2″ × 8″

4.   Exterior wall studs—2″ × 6″

5.   Interior studs—2″ × 4″; where plumbing is to pass vertically, use 2″ × 6″ studs

6.   Carrying beams—4 each 2″ × 10″, bolted together every 48″, with nails in between

7.   Door and window headers—2″ × 10″

C.   All subfloor, wall, and roof sheathing shall be as specified:

1.   Subfloor sheathing—$\frac{3}{4}$″ T&G exterior-glue U.L. plywood
2.   Wall and roof sheathing—$\frac{1}{2}$″ CDX exterior-glue plywood

CDX panels are not designed to withstand prolonged exposure to weather and the elements; they lack A-grade or B-grade veneers. They should be covered as soon as possible to prevent excessive moisture absorption. The moisture raises the grain and causes checking and other minor surface problems.

D.   Fasten the subfloor sheathing with an adhesive and 8-penny annular-ring nails spaced 12″ on center, except for the butt ends. (The butt ends are to be fastened every 8″.) When fastening the subfloor sheathing, leave a $\frac{1}{8}$″ space at all edges and end joints. Important: Do not hammer the sheathing to close the joints. Cut the end of an adhesive caulking gun at a 45-degree angle and position the gun perpendicular to the joist. Apply a $\frac{3}{16}$″ bead in a continuous line except where the butt ends of the plywood meet. Apply two beads for each butt end. (The adhesive prevents squeaky floors.) Recommended adhesive manufacturers are H. B. Fuller Company; DAP, Inc.; and Miracle Adhesive Company.

E.   Bridging will be required. It shall consist of either commercial metal bridging or strapping cut to length and toenailed to the joists.

F.   Wall sheathing shall be fastened with 6- or 8-penny common nails, 8″ on center, except for the butt ends. The butt ends are to be fastened every 6″.

G.   For 2″ × 6″ exterior wall openings for door and window headers, use box beams, prefabricated and formed by gluing and nailing plywood web to the headers.

H.   All rooms are to be strapped with 1″ × 3″ strapping, placed 16″ on center, each secured with two 6-penny or 8-penny nails.

I.   The ridge boards are 2″ × 10″. At the ridge, leave 1.5″ air space between the ridge boards and the last sheet of plywood, both sides of the ridge, to provide an opening for the ridge vent.

J.   Nails for installation of doors, window, and trim shall be as follows:

1.   Exterior door frames shall be installed with 10-penny galvanized finish nails, countersunk $\frac{1}{8}$″, the heads covered with oil-based putty.
2.   All window units shall be installed with 8-penny galvanized finish nails, set and puttied.
3.   All exterior trim shall be installed with 8-penny galvanized finish nails, set and puttied.
4.   All interior door frames and trim shall be installed with 8-penny finish nails, set and puttied.

The shoe is a horizontal timber, to which the vertical wall timbers are fastened.

K.   The shoe for the 2″ × 4″ basement-carrying partition shall be pressure treated. It is to be installed on 15# asphalt saturated felt, secured 6″ up on both sides. (The wood will be less apt to absorb dampness.)

Note that ramps are specified. This allows access by all people, whether or not they are physically handicapped.

L.   All lumber used for the ramps and porches shall be pressure treated with chromated copper arsenate (CCA), sealed and/or stained. All nails for the ramps and porches shall be galvanized.

ROOF

A.   Roof shingles shall be GAF Timberline Series shingles, 30-year Timberline in "Pewter Grey" blend, Class A. The ridge vent shall be the Shinglevent II as manufactured by Air Vent, Inc., or an equal vent system.

B.   Roof sheathing is to be installed in accordance with the manufacturer's recommendation, leaving $\frac{1}{8}''$ between each butt end, with 8-penny common nails at 12″ on center, except at the butt ends. The butt ends are to be fastened every 8″.

Each roof sheathing manufacturer recommends a different application technique that depends upon temperature, wetness, and lumber treatment.

C.   The roof covering is to be installed within seven (7) working days after installation of the roof sheathing. When fastening the sheathing, leave a $\frac{1}{8}''$ space at all edges and end joints. Important: Do not hammer the sheathing to close the joints.

D.   Install roofing as soon as practical to prevent roof sheathing from absorbing moisture, or being affected by rain or sun. Install the roof system in conformance with the Asphalt Roofing Manufacturers Association (ARMA) recommendations that require $\frac{1}{8}''$ spacing at edges of all plywood and 15-pound asphalt paper conforming to ASTM specification D4869 or D226. Their specified installation sequence is as follows:

Snow collects on a roof during an extended period of continuous below-freezing, snowy weather. The heat from the interior of the house, combined with sunshine, causes the snow on the upper part of the roof to melt. The cold water runs down the roof. This cold water, now on the lower part (overhang) of the roof, freezes and causes an ice dam. Additional cold water that is running down and is not yet frozen is blocked by the ice dam. It works up under the shingles and into the interior of the building, causing ceiling and wall damage.

1.   Install the eave drip edge first.
2.   Install the ice-and-water shield over the drip edge at the eaves and up the roof 3′ above the exterior wall line of the building.
3.   Attach the 15# asphalt saturated felt to the entire roof, overlapping the ice-and-water shield a minimum of 6″. Apply roof cement between the ice-and-water shield and the asphalt felt.
4.   Install the drip edge over the asphalt felt.
5.   Use aluminum nails with an aluminum drip edge, or galvanized nails with a galvanized drip edge.

E.   When roof shingles are installed, nails must be located below the adhesive line. (Nails located above the adhesive line affect the shingle wind resistance, allowing snow or roof ice to be driven under the shingles, thus causing water backup.)

WINDOW AND DOOR SCHEDULE

A.   Windows and doors shall be installed in accordance with the manufacturer's recommendation. Fifteen-pound asphalt saturated felt shall be installed under all windows, and extend 4″ beyond the window. (Typar® wrap may be used; it will eliminate the need for the asphalt felt.) A $\frac{3}{4}''$ space shall be left between the sides and tops of the door and window frames, and the studs and headers, for eventual insulating-foam application. (This is because $\frac{1}{2}''$ is too tight. It is easier to fill the $\frac{3}{4}''$ space with the nozzle of a standard insulation gun.) Install aluminum header flashing above all exterior doors and windows.

B.   Door and window wood surfaces shall be stained with client's choice of stain; finished with two coats of (nonpetroleum-based) polyurethane. Exterior doors shall be finished with color of client's choice.

C.   The window schedule is as follows (all windows and doors shall include extension jams and screens):

Andersen, 4 each of #2442

Andersen, 1 each of #5042

Andersen, 6 each of #3042

Andersen, 1 each of #CR23

Andersen, 1 each of #3032

Door widths were selected to provide wheelchair access.

D.   The door schedule is as follows:

Exterior Pre-Hung Doors—3'0″ × 6'8″ frame opening

1.   Two Brosco BE-70 Steel Doors, steel-insulated, with foam-insulated cores
2.   Hardware: double-cylinder dead-bolt locks, brass handles, and key-in knobs

Interior (Pre-Hung) Doors—3 each, 3'0″ × 6'8″ Morgan six-panel pine, sealed on all six sides (two coats) to reduce moisture absorption

EXTERIOR TRIM AND SIDING

A.   The entire house shall be enveloped with Typar®, manufactured by W. R. Grace.

B.   The exterior house trim—rakeboard, fascia, and plancia—shall be covered with prefinished vinyl or aluminum trim cover. A continuous white aluminum/vinyl louver shall be installed on the plancia board, approximately 2″ from the fascia board.

Suppliers of caulking may no longer offer 50-year-life; "or equivalent" is often specified.

C.   The exterior siding shall be 16″ S&R (squared & rebutted) white cedar shingles, installed 5″ to weather. The corner board shall be of rough white cedar or fir to match the shingles. The shingles and corner board are to be coated with bleaching oil supplied by the client. At the areas where the shingles butt against the windows, the doors and rakeboard shall be caulked with a 50-year-life white caulking. The shingles shall be installed with 5-penny galvanized box nails. The contractor shall provide an extra bundle of shingles to the client for future repair purposes.

PLUMBING AND HEATING

Internal plumbing is within internal walls to decrease the possibility of pipes freezing.

A.   The water line from the water main to the meter within the activity center will include a repeater. It is to be purchased and installed by the contractor. The Bedford Municipal Water Company contact (Mr. Alfred Horch) requests that the contractor contact him before any digging commences and before any materials are purchased in order to ensure compatibility with the water-main materials. At the water-line entrance to the house, there are to be a shutoff valve, the water meter, and a back-flow prevention device, in that sequence. The pipe from the water main to the house must be buried at a depth and in a manner acceptable to the water company. The materials costs are to be included in the contract.

B.   The sewer line is to be installed according to the attached document titled: Town of Bedford Sewer Tie-In Application. The pitch of the sewer line from the house to the town sewer line is to be at least $\frac{1}{4}$″ per foot.

C.   Internal plumbing is to be installed in conformance with the requirements of the Town of Bedford code. Waste lines are to consist of 4″ PVC pipe. All internal water lines are to be installed with type L copper. All plumbing is to be pressure-treated and examined by the Town of Bedford plumbing inspector prior to approval. All valves are to be tagged, stating their purpose (function).

D.   The following bathroom fixtures are to be purchased and installed by the contractor unless otherwise noted:

1.   Toilet: Kohler, Wellworth Lite, 1.5-gallon flush, K-3420-EB; color—Heron blue
2.   Toilet Seat Cover: Kohler Lustra
3.   Bathroom Sink:

Panel Vanity—Merrilat Industries, Model #782430; 30″ × 21″; 1″ door, 2 drawers

Vitreous China Top and Drop-In Lavatory—Merrilat Industries, Model #4638; 31″ × 22'1$\frac{1}{2}$″—white

4. Contemporary Mirrored Medicine Cabinet—Merrilat Industries, Model #7714305; 30″ × 26″

5. Faucet for Bathroom Sink—UPC#0-34449-00663: Delta Chrome Model 575PHDF with (single) lever handle and pop-up drain

E.   The following kitchenette fixtures shall be purchased and installed by the contractor:

1. Sink, Kohler; K3228, 18-gauge insulated undercoat

2. Faucet, Kohler; K15170AP plus basket strainer Kohler K8801-PC

F.   The contractor shall install a standard 2′ × 6′ countertop for the kitchenette sink. The contractor shall cut a hole in the countertop suitable for installing the sink and shall install the kitchen sink.

Paragraph G is included should the town of Bedford later decide that the activity center is not to be used during the winter months.

G.   All waste lines and water lines are to be installed with a pitch to permit the water to drain to the basement without entrapment of any water in the lines for winter-season drainage.

H.   Two exterior freeze-resistant sill cocks are to be installed. These sill cocks shall be pitched to drain to the outside. One is to be located on the south side of the center and one is to be located on the north side near the door in locations satisfactory to the client.

I.   The contractor is to purchase and install an energy-efficient 40-gallon electric hot-water heater with a hand-operated shutoff valve and having a ten-year warranty. A brick support is to be constructed to raise the water heater 4″ above the floor. The drain in the basement floor is to be under the faucet of the water heater so it will remove water from seasonal tank drainage, accidental spills, or leaks from the water heater.

J.   The contractor is to install the Boston Gas–donated hot-water, gas-fired furnace and associated radiators under the guidance of Boston Gas Company personnel.

ELECTRICAL

The distribution panel must contain a lock because children will be present. It must also be low enough so an attendant in a wheelchair can access the panel.

A.   The contractor is to supply and install an electric service to be 100A@230V with one lockable distribution panel on the main floor to be provided by the contractor. The distribution panel is to contain a master circuit breaker and space for 20 other circuit breakers. Power-distribution panels: the bottom edge is to be 42″ from the floor. The electric power is to enter the house via an underground-feed (conduit) service entrance on the south side—2′ from the southwest corner. The KWHR meter is to be installed externally.

B.   The contractor is to supply and install the following fixtures for the main floor bathroom:

1. Radiant Wall Heater, NuTone—Model #9358

2. Fan/Light NuTone—Model #9965; 4″ duct required

WALLS, FLOORS, AND CEILINGS

Bathroom

This interior insulation provides a bathroom that is quiet.

A.   Install ½″ moisture-resistant Sheetrock™ on the bathroom walls and ceiling. Insulation for the interior walls shall be R-19 fiberglass with a kraft-based or foil-based vapor barrier.

B.   The wall covering for the bathroom shall be American Olean Crystalline Azure Blue 4.25″ × 4.25″ ceramic tile. The tile is to be installed from the floor to a height of 46″ to 48″.

The area above the tile shall be wallpaper: Breakfast in the Country Book, Lorraine Stripe #2382718, price code B, Fashion.

C. The floor covering for the bathroom shall be Armstrong Sundial Solarian #66174.

D. The bathroom ceiling shall be a suspended ceiling with 12″ × 12″ Armstrong Vinyl-coated Temlok Ceiling Tile; #225 Glenwood.

Remainder of Activity Center

E. Insulation for all exterior walls shall be R-19 fiberglass with a kraft-based or foil-based vapor barrier.

F. Ceilings shall consist of $^3/_8$″ Sheetrock™ with a haircoat of plaster and a glazed finished coat.

G. Walls shall consist of $^3/_8$″ Sheetrock™ with taped and coated joints, sanded to a smooth finish. The walls shall be covered with one coat of primer and two coats of wall paint to be selected and supplied by the client.

H. Floors shall consist of $^5/_8$″ CD plywood covered with 12″ square linoleum tile. The tile is to be supplied by the client.

I. Window, door trim, and baseboards will consist of soft pine with a natural finish. The pine will be supplied by the client.

LOCKS

The volunteers who manage the activity center will be able to enter via the handset locks. The dead-bolt locks permit a town employee responsible for the building to prevent entry during unauthorized times.

A. All internal-door lock sets are to be provided by the client and installed by the contractor.

B. The two lock sets in the exterior doors will each consist of a dead-bolt lock and a handset lock. The two dead-bolt locks are to be keyed alike. The two handset locks are to be keyed alike. The dead-bolt locks are to be keyed differently from the handset locks.

LANDSCAPING, DRIVEWAY, RAMP, AND WALKS

Gutters and downspouts are thereby eliminated, reducing maintenance requirements.

A. The area from the foundation, out to 3′ from the foundation on the north and south (drip edge) sides, is to be excavated to a depth of $5^1/_2$″. Pressure-treated 2″ × 6″ lumber shall be installed on the north and south sides of the foundation, parallel to the foundation at the edge of the excavated area and level with the ground. (See the document "Finish Grade Information.") The lumber is to be secured by stakes in the ground. The remaining excavated volume is to be filled with processed gravel.

B. An approximately 25′ × 26′ parking area shall be constructed in an area 10′ from the street and 10′ from the north property line via an 8′-wide driveway from the street. The size of the grassed-over parking area is to be approximately 25′ from north to south and 26′ from east to west, with its north edge approximately starting along the line of the south side of the activity center. Pressure-treated 2″ × 4″ lumber (or 6″ × 8″ railroad ties), secured by stakes, shall be installed around the three nonroad sides of the rectangle to contain the grassed-over parking area. Client shall supply the grass seed.

C. A 3′-wide walk shall be installed from the parking area to the ramp using either processed gravel or granite dust. The ramp is described in the drawings.

APPENDIX D
Bedford Activity Center Contract

This contract represents an agreement between the client (who is acting for the selectmen of the Town of Bedford, Massachusetts)

James T. Williams, Superintendent of Public Works
Town of Bedford
44 Mudge Way
Bedford, MA 01730

and the contractor (who is acting for Northeastern University)

Thomas E. Hulbert, Associate Professor of Industrial Engineering
Northeastern University
360 Huntington Avenue, Room 467 Snell Engineering
Boston, MA 02115-5096.

The project consists of constructing an activity center and landscaping the associated grounds within thirty (30) feet of the activity center at

The Bedford Common, Southwest Corner
The Great Road
Bedford, Massachusetts 01730.

Normally, the client supplies drawings via an architect to the contractor.

The contractor shall supply to the client one (1) set of reproducible drawings of the activity center (sepias, also known as brownlines) within seven (7) days of contract signing.

During construction and landscaping of the new structure, the contractor will supply promptly (as appropriate) either videotapes or photographs (that the client will retain) of the status of site preparation, new-structure construction, changes that circumstances may require, and landscaping.

Cost of This Contract is: $20 000, fixed price.

Signed and Accepted: _____; date: _____
    (James T. Williams, client's representative)

The remaining pages are initialed by the client and contractor.

Signed and Accepted: _____; date: _____
    (Thomas E. Hulbert, contractor's representative)

A list of contract documents is given on page 2 of this contract.

Client's Initials: _____; Date: _____; Contractor's Initials: _____; Date: _____

Contract—page 2 of 3

Contract Documents

> 1 each Specification for the Town of Bedford, Massachusetts: "An Activity Center," dated June 6, 1996
> 1 each Finish-Grade information, dated May 17, 1996
> 1 each Grade Information (enlarged), dated May 28, 1996
> 1 each Town of Bedford Sewer Tie-In Application dated 11-7-95 and permit dated 2/12/96
> 1 each building permit, dated March 7, 1996

and

> 1 each Task Description for Construction of Bedford Activity Center
> 1 each set of drawings, dated 1996 June 11, titled: Bedford Activity Center,

consisting of

> A1 — Site Plan
> A2 — Foundation Plan
> A3 — Basement Plan
> A4 — Main Floor Plan
> A5 — Cross-Section
> A6 — Long Section
> A7 — East Elevation & West Elevation
> A8 — North Elevation & South Elevation
> A9 — Door & Window Schedules

Any agreed-upon Contract Change Notice to these attachments shall be endorsed by both the client's representative and the contractor's representative.

Warranties

In addition to warranties given by vendors, the contractor warrants the construction and materials for a minimum of one year.

Client's Initials: _____; Date: _____; Contractor's Initials: _____; Date: _____

Contract—page 3 of 3

Schedule of Performance and Progress Payments
The "start" date is: 1996 March 1
*   completion of foundation forms prior to pouring of concrete
    payment: 10% on completion, expected date: 1996 April 1
*   pouring of foundation concrete (James T. Williams will witness slump test)
    payment: 10% on completion, expected date: 1996 April 8
*   framing, roof, and "boarding in" prior to shingling
    payment: 15% on completion, expected date: 1996 April 29
*   pouring of basement-floor concrete (Mr. Williams will witness slump test)
    payment: 5% on completion, expected date: 1996 May 6
*   shingling of roof and sides, and installing exterior trim
    payment: 15% on completion, expected date: 1996 May 20
*   installation of water, natural gas, electricity, and telephone connections
    payment: 5% on completion, expected date: 1996 June 17
*   completion of tie-in to Town of Bedford water and sewer system
    payment: 5% on completion, expected date: 1996 July 15
*   completion of bathroom and kitchenette plumbing
    payment: 5% on completion, expected date: 1996 August 19
*   completion of interior finish (ceilings, walls, and floors)
    payment: 10% on completion, expected date: 1996 September 3
*   completion of exterior landscaping and parking area
    payment: 10% on completion, expected date: 1996 September 9
*   final inspection and acceptance
    payment 10% on completion, expected date: 1996 September 16

A penalty of $2 000 will be assessed the contractor if work is not completed by 1996 September 23rd.

Compensation in the form of a check shall be mailed to the contractor upon completion of work, or agreed-to partial work, within seven calendar days from receipt of an approved invoice/bill.

Progress and funds shall be monitored by two selectmen of the Town of Bedford and funds released only by checks signed by them.

Client's Initials: _____; Date: _____; Contractor's Initials: _____; Date: _____

For the case study, total payment occurred on 1996 March 1; the school's internal release of funds is noted here.

APPENDIX E

Bedford Activity Center Contract Change Notice (CCN)

The following is an example of a change notice used to communicate, in writing, the desired changes to a contract. The contract is assumed to be already signed and, therefore, in effect. Many companies display the sign "AVOID VERBAL ORDERS" to remind their employees to document changes in writing in order to protect all involved parties from later misunderstandings.

1996 July 14

TODAY'S DATE	CONTRACT CHANGES	PRICE INCREASE	PRICE DECREASE
96 July 03	Spec page 242 ¶ C. & Drawing page A11, Window Schedule: change Anderson Double Hung #B, D, E, H, L, M, N, O, P from #3042 to #DC3042 (tilt-in feature)	$400	
96 July 03	Spec page 239, Cast-in-Place Concrete ¶ B & Drawings, pages A2, A3, A6, A7, A11: change 5" thick nonreinforced to 4" thick reinforced with fiberglass mesh		$200
96 July 03	Drawing page A6: Change Simpson TP gussets to plywood-scrap gussets on both sides of all inner trusses		no charge
96 July 03	Spec page 238 ¶ A—Site Utilities— include a spare pipe for cable TV	$20	

Signed: _____; _____
Contractor   Client

Date: _____   Date: _____

APPENDIX F
Records and Their Interactions

The goal of this appendix is to introduce the reader to record design and maintenance. This appendix will also focus upon the content of records, how they interrelate, and how they are initiated, maintained, and accessed via a computer. Specialized texts are available for readers who are interested in learning more about records—their content and design. Sources include Carrithers and Weinwurm (1967), Graham (1982), Isaak (1994), and Kaimann (1973).

Recordkeeping started more than 4000 years ago. Records were initially kept on stone, then on parchment, and later on paper.

Paper records are known as hard copy. They require large amounts of space in storage cabinets and file drawers. As the cost of computer memory decreased and the speed of computer processing increased, more records were both generated and stored in computer memory. The cost of recordkeeping decreased and the ease of access increased.

As programming became more sophisticated, the concept of records changed. Computer speed, memory costs, records access, and computation flexibility have allowed for greater record interaction. This appendix is an overview of the progress achieved in records and their design.

As projects became larger and more complex, it was necessary to provide information regarding a project's initial estimates, its latest status, and the present plans. In the past, documents of this type were prepared and maintained manually—by hand. Today, with the assistance of inexpensive and flexible computers, these documents are prepared and maintained via computers that use special programs.

SECTION F.1   TYPES OF COMPUTER RECORDS

The word **record** has two meanings. It is the name of a document that contains information worth developing, distributing, and maintaining. It is also the name of a portion of a computer file. A computer record is information that is stored electronically. This record may contain many **fields** of information—a concept that will be discussed later. A record or field may be structured (fixed) or unstructured (open ended).

A **structured record or field** is a set of categories of related data. Every set of fields is identical in content, arrangement, and length. For a structured field, the maximum length of each item within that field must be defined in advance. For example, a structured field that can be used for both generation of mailing labels and quick access to a person's telephone number might contain a total of four items. See figure F.1.

IDENTIFICATION ITEM—9 characters

These nine characters are allotted for each individual's social security number, which is a unique descriptor assigned by the U.S. federal government. The Federal Privacy Act, many states, and some other organizations prefer to develop their own numbering system rather than use social security numbers.

Figure F.1 Structured field within a record

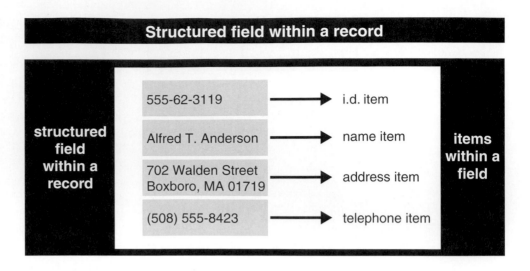

NAME ITEM—25 characters

These twenty-five characters were determined from an extensive study of name lengths. Longer family names are often truncated (shortened); exceptions may be stored in a separate, unstructured file whose location then becomes part of this structured file.

ADDRESS ITEM, LINE #A—27 characters

These twenty-seven characters were determined by the length of a given mailing label. Mailing labels are available in many lengths and widths.

ADDRESS ITEM, LINE #B—27 characters

These twenty-seven characters were also determined by the length of the mailing label.

TELEPHONE ITEM—14 characters

For the United States, whose telephone number formats were established by AT&T many years ago, the fourteen characters are allotted to

  area code—3 characters

  telephone exchange—3 characters

  person's number within that exchange—4 characters

  extension number if applicable—4 characters maximum

Figure F.2 Unstructured field within a record

Unstructured field within a record

555-62-3119
Alfred T. Anderson
  84 Jun 04
          BSMT– Northeastern University
  84 Jun 12 to 88 Sep 15
          Mechanical Technologist
          D'Annolfo Construction Co.
  88 Sep 29 to 92 Jan 05
          Senior Mechanical Technologist
          Buttaro Development Corporation
  92 Jan 19 to present
          Project Planner
          Bob & Norm's Student Training, Inc.

Within a structured field, all items are in the same location with respect to the start of the field. Therefore, programs can be written that search a given portion of that field. They can compare and prepare a document that, as examples, lists all persons with the same street address, zip code, or telephone number.

Sometimes three-digit country codes are included. Thus, the maximum number of characters for this entire record is 102 or 105. Memory space is allotted for the maximum number of structured records (or fields) to be stored for later use.

An unstructured information record or field is a set of categories of data related to one another. Every item within each field may be different in content, arrangement, and length. For an unstructured field, the maximum length of the total field must be defined in advance in order that memory locations and size may be allotted. Unstructured fields vary in length because some descriptors will require more storage space than others. Unstructured fields are searched using key words, phrases, or numbers. The use of unstructured records and fields allows for greater record-design flexibility. An example of an unstructured field would be the resume of Alfred T. Anderson, shown in figure F.2.

Structured fields are usually of the same length. Such records may be inefficient because fields may contain blank spaces to allow for the longest acceptable word or number. However, because the location of the beginning of each item occurs at the same position within its field, that particular item may be quickly accessed and used for other purposes.

SECTION F.2   RECORD INTERRELATIONS

We have been examining the interrelation of various items and fields within a record. A **database** is a group of interrelated records.

Databases are only as accurate as the data submitted to them. The phrase "garbage in, garbage out" aptly describes the relationship of input and output when inaccurate data are included in a database. Financial auditors periodically examine the contents and historical records of a database. They want to assure corporate executives that the information entered from the hard copy is the same as the data within the database and that the database has not been altered without proper authorization.

Here are some basic definitions relating to databases, using personnel information as an example:

- Database—a group of records that describes several persons, places, or things
    Personnel example: the persons who work for the company
- Record—a group of fields that describes one person, place, or thing
    Personnel example: employment history, attendance reports
- Field—a group of items that contains information regarding one person, place, or thing
    Personnel example: (full) name, address, and telephone number
- Item—a piece of information that is to be stored for later processing
    Personnel example: last name, first name, middle initial

The database hierarchy is outlined in figure F.3.

Thus, a database consists of many records, as displayed in figure F.3. A record consists of many fields. A field consists of many items. An item consists of many characters or digits. A character or digit consists of many computer bits.

Figure F.3 Typical hierarchy of an electronic database

Typical hierarchy of an electronic database

database–(most significant level)
    record
        field
           item
              character or digit
                computer bit (least significant level)

Figure F.4 Information flow of time charges into database

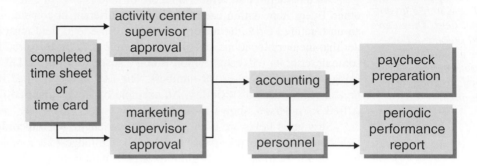

Information flow of time charges into database

Memory sizes are grouped into **bytes.** One byte consists of eight bits. The quantity of computer bytes is now used to describe the size of a computer memory. As an example, a 3.5″ (9 cm) floppy computer disk can store 1.3 million bytes of usable data, expressed as 1.3 megabytes.

The content of a record depends on both the intended use of that information and the design of the forms used for record entry and display. The display may be on a computer screen display or through a printed hard copy.

Examine the information flow for the time charges of Alfred T. Anderson. Assume that Alfred T. Anderson is a salaried employee. He is working part of the time assisting Marketing in searching for new business and part of the time working on the activity center project. Alfred is issued two charge numbers—one by Marketing and the other by the manager of the activity center project. The typical flow of this information, whether hard copy or computer data, is shown in figure F.4.

All employees usually submit attendance data weekly, either using a time sheet if salaried or via a time-clock-generated card if hourly. For each different project upon which Alfred worked, he provides the time that he worked on that project on his time sheet or card. He has two entries. Each is identified with the assigned charge number and the number of hours worked for each is noted. This hard-copy information is then approved by the appropriate supervisor(s) and entered into the appropriate parts of the database.

A supervisor may sign the approval via a computer "pad." This saves hard-copy paperwork and allows others to verify the signature at a later time.

Alfred is working part of the time on the activity center project. The head of that project will want to examine and verify Alfred's charges to the project. Alfred is working part of the time searching for new business for the Marketing Department. The person responsible for that activity will want to examine and approve Alfred's charges to this effort.

Note that two other departments have a special need for Alfred's time charges:

- Accounting, so that it may record his time charges and prepare a paycheck for him
- Personnel, so that they may determine if he has used his vacation or sick leave and also so that they may accumulate data for his periodic performance report

Accounting personnel must have access to the payroll portion of Alfred's record. However, they have no need to know about his personal nonpayroll data. Thus, restricted access codes are used to prevent unauthorized persons from gaining access to other files and items within Alfred's record.

The Personnel Department will want to receive, monitor, and retain information regarding Alfred's time charges and performance. Members of the Personnel Department are often responsible for obtaining all performance inputs so Alfred may be evaluated for a merit raise and/or promotion.

The Marketing Department will need to know how much time and money is being charged to its budget. Alfred's time sheet or time card provides this information. Alfred's information is entered into the database, and data information from the files of Alfred and all others who have assisted the Marketing Department can be extracted from the database.

The activity center manager, Roger, will want to know how much time and money is being charged against his effort. Manuel will have prepared the form for presenting the data desired by Roger. Those persons completing this document must have access to the files of Alfred and all others who have assisted in this project.

Figure F.5 Typical structure
of a relational database

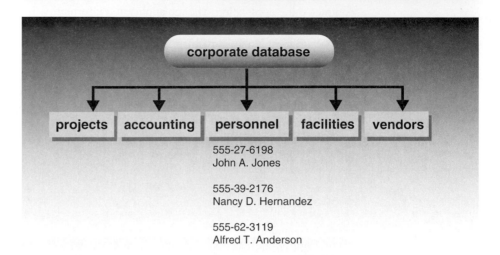

Typical structure of a relational database

The database may be a relational or interrelated database. Therefore it can be shared by several groups or organizations, such as shown in figure F.5.

Each group or organization may have total access or limited access to data, depending on its needs. Modification of the database must be further restricted to avoid improper alteration of records.

Database activities include data gathering, manipulation, storage, and retrieval. An interrelated database (figure F.5) may consist of the following types of records:

Projects
    External (such as activity center)
    Internal (such as research)
Accounting
    Project Costs
    Payables and Receivables
    Department Costs
    Balance Sheets
    Profit and Loss Statements
Personnel
    Employees
    Job Descriptions
    Employment Agencies
Facilities
    Buildings
    Furnishings and Fixtures
    Paved Areas
    Landscaped Areas
    Maintenance
Vendors
    Office Supplies
    Electronic Parts
    Computer Hardware and Software
    Machine Tools
    Assembly Equipment
    Security

Within the storage medium of a computer, there is an index of key words or groups of digits. These words or digits are used to locate records, such as personnel records.

As an example, a new personnel record may be initiated from hard copy, such as from a completed job application. The job application information is arranged in a form so that the data can be readily processed by and stored within a computer.

Data storage sizing involves estimating the number of records and the length of each record. Data retrieval requires knowing how each type of record has been designed so access to that record is simple and easy to comprehend. The computer data storage subsystem is indexed so records, or portions of records, may be easily accessed by authorized departments. Also, each record, whether structured or unstructured, must contain locations that always contain the same type of information. Figure F.6 indicates a flow of information related to the preparation of a paycheck and the associated payroll data.

Figure F.6   Flow of payroll information

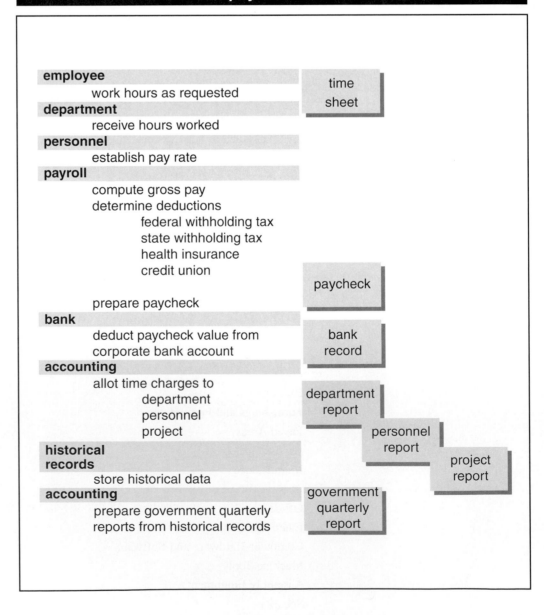

A database is designed so that the data contained can be used to formulate a variety of documents. Thus, interrelated information can be extracted and turned into documents. Each document will have a standard format and layout that allows information to be transported from the database into the correct portion of that document. Typical documents include the following:

Current and Potential Vendors: capability and prices

Office Supply Sources: supplies and prices

Personnel: staff and consultants

Occupancy: rent, utilities, and maintenance

Insurance and Taxes

    Business Liability Insurance

    Workers' Compensation Insurance

    Unemployment Taxes

Communications

    Telephone

    Copying

    Fax

Travel

    Transportation

    Lodging

    Meals

Inventory

    Construction Materials

    Office Supplies

    Furnishings and Fixtures

    Equipment

Typical information to be accessed includes the following:

Per Person (This also includes information for Payroll.)

    Labor Rate

    FICA Rate

    Federal (IRS) Tax Rate

    State (DOR) Tax Rate

    Local Tax Rate

    Fringe Benefits, such as health insurance

Per Project

    Tasks versus Months Schedule

    Labor versus Months Schedule, and Labor Rates for

        Foreman

        Carpenter

        Masons

        Electrical

For the Corporation

   Cost Control Records

      Estimates versus Actuals

      Payroll: salaries and wages

      Inventory

         Materials

         Supplies

         Furniture

         Equipment

      Taxes

         Federal

         State

         Local

These categories will vary from one corporation to another, and from one database to another. They may be displayed on a computer screen or converted to hard copy.

SECTION F.3   DOCUMENT DESIGN

Computer programs can be designed by a programmer to extract information from an interrelated database and place it into documents. See figure F.7. The programmer designs a Payroll document and also designs a Project Report document. The programmer writes a program for the Payroll Department that indicates the following: FICA information is in location 002, Hourly Rate is in location 003, Hours Worked is in location 004, and Gross Pay is in location 005. The program will automatically extract this information from an employee's record, such as Alfred's record.

For the Activity Center Project Report, the programmer writes a program that indicates the following:

1. FICA information is in location 002 of an employee's record.
2. (Activity Center) Hours Worked is in location 006.
3. (Activity Center) Gross Pay is in location 008.

The program will automatically extract this information from Alfred's record. For Nancy Hernandez, John A. Jones, and others, the same program can be repeated to extract the same information.

In practice, the record for Alfred T. Anderson and all other employees will contain many more items and fields than are shown in figure F.7. As an example, at the time each employee is hired, entries and fields might be the following:

   Employment Application

   Interviewer's Critique

   Reference Checks

      Payroll Authorization and Deductions

      Gross Pay (hourly or salary)

      FICA (Federal Insurance Contributions Act), more commonly known as social security

      Withholding: federal, state, and perhaps local

      Health Insurance: medical and dental

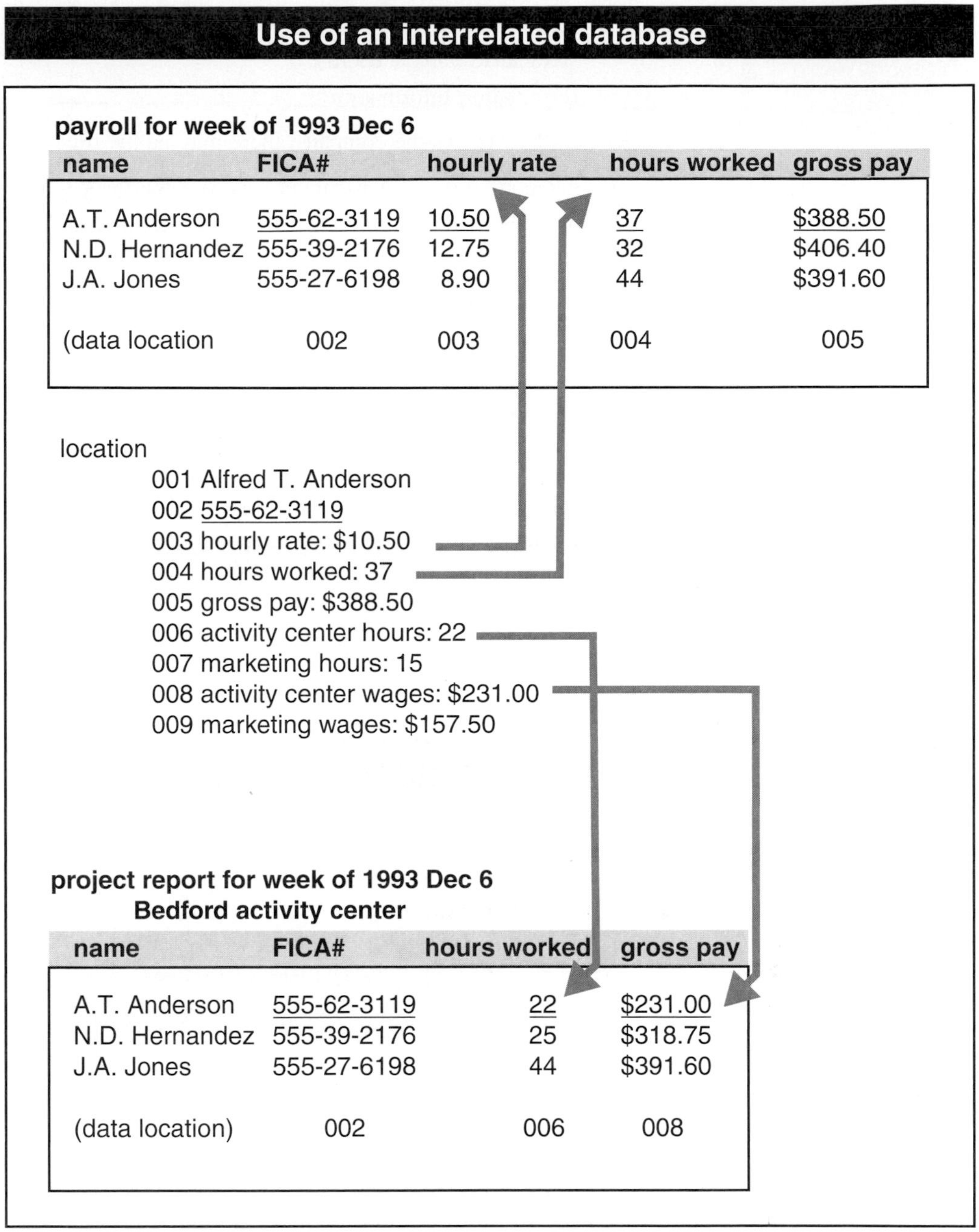

Figure F.7 Use of an interrelated database

Use of an interrelated database

payroll for week of 1993 Dec 6

name	FICA#	hourly rate	hours worked	gross pay
A.T. Anderson	555-62-3119	10.50	37	$388.50
N.D. Hernandez	555-39-2176	12.75	32	$406.40
J.A. Jones	555-27-6198	8.90	44	$391.60
(data location	002	003	004	005

location
      001 Alfred T. Anderson
      002 555-62-3119
      003 hourly rate: $10.50
      004 hours worked: 37
      005 gross pay: $388.50
      006 activity center hours: 22
      007 marketing hours: 15
      008 activity center wages: $231.00
      009 marketing wages: $157.50

project report for week of 1993 Dec 6
Bedford activity center

name	FICA#	hours worked	gross pay
A.T. Anderson	555-62-3119	22	$231.00
N.D. Hernandez	555-39-2176	25	$318.75
J.A. Jones	555-27-6198	44	$391.60
(data location)	002	006	008

During their employment, the records might be the following:

    Periodic Evaluations: performance reviews

    Attendance Information

        Time Clock or Sheet (date, hours worked, project or task number)

        Vacation Days

        Personal Days

        Other Time Off: with and without pay (direct and indirect charges)

    Salary and Salary Change Data

        Payroll Change Notice (including deduction changes)

            Withholding: number of exemptions

            Credit union: deposits

            Health benefit changes

Accident Report

Medical Report

Termination Information

When Alfred leaves the company, there may be the following:

1. A report from the department head or supervisor to the Personnel Department

2. A closure notice to the Payroll Department and to Alfred's department head

Project managers often need to access portions of the above information. Their access may be direct, via a computer network, or through the Personnel Department, depending upon company policy.

SECTION F.4   CORPORATE DATABASE REQUIREMENTS

An interactive corporate database requires the ability to retrieve and manipulate data from the company's customers, vendors, and its own corporate departments. These data (shown in figure F.8) will be in identified locations similar to those in the previous personnel example.

The overall corporate storage requirements include all those records developed and kept by the departments indicated in figure F.9. As mentioned earlier, access is limited to those persons who are authorized to both view and modify these records.

Figure F.8 Corporate interactive storage

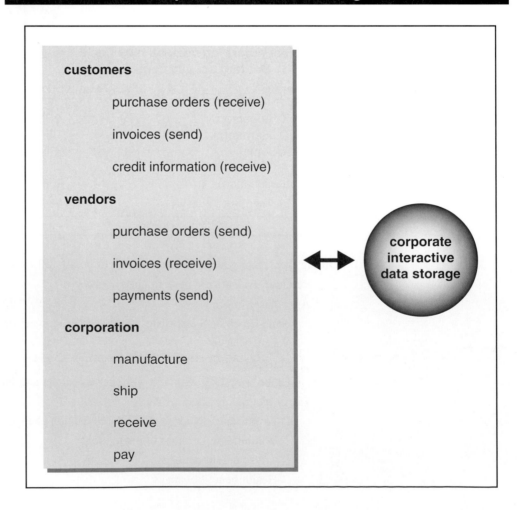

Figure F.9 Corporate
overall storage
requirements

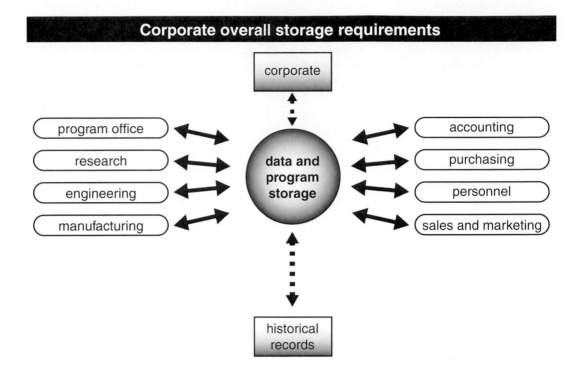

Note that historical records—archives—are also kept and cannot be modified: They can only be retrieved. A well-trained auditor can also determine if someone has altered records in an inappropriate manner. This auditor can often trace the alteration to the computer terminal on which the alteration was entered.

Computer records can be disseminated to various locations within a corporation and to other corporations with which the corporation interacts. Within a given building or adjacent building, Local Area Networks (LANs), such as Ethernet, Token Ring, and FDDI, can be utilized. Where the locations are far apart, the Wide Area Networks (WANs), such as X.25, SDLC, SMDS, B-ISDN, and ATM, can be utilized. Popular protocols that allow database interaction include FDDI, X.25, and X.400.

A National Information Infrastructure (NII), referred to in the nontechnical press as the Information Superhighway, is being designed and tested. This infrastructure will span a multitude of standards-development organizations on a worldwide basis. The challenge is to work together at all levels to provide a coherent environment for the NII. The standards that are incorporated into the NII will eventually link every household and desktop in the nation and, later, in the world.

APPENDIX G
System-Level Specifications

GENERAL

The general content and hierarchy of a system-level specification is given in section 8.1, figure 8.2, and is repeated here.

- Scope of Specification
- Applicable Documents
- Requirements

  System Definition

  System Characteristics

  Design and Construction

  Documentation

  Logistics

  Personnel and Training

  Precedence

- Quality Assurance Provisions
- Preparation for Delivery
- Notes and Appendixes

This appendix will focus on the "System Definition" and "System Characteristics" sections within the Requirements paragraph. This section is the more technical portion of a system-level specification. Typical content of these sections is given below for each of the following fields of interest:

Aeronautical

Architectural

Biomedical and Biotechnology

Chemical

Civil

Communications

Computer

Electrical and Electronic

Environmental

Hazardous Waste

Industrial

Manufacturing

Marine and Marine Science

Materials

Mechanical
Nuclear
Optics
Systems

AERONAUTICAL

System Definition
System Characteristics

ARCHITECTURAL

System Definition

For a given piece of land, devise a compatible and attractive family of buildings containing no more than five floors. Design one structure for use by retail stores, a second structure as an office building, and a third structure as a warehouse. Include in the design a parking area and landscaping that causes the entire family of buildings to look pleasing to a passing observer.

System Characteristics

The dimensions of each building shall not exceed: 50′ in height, 60′ in width, and 30′ from the front to the rear of the building. Local building codes will apply with regard to frontage requirements.

The first structure shall contain second-floor-and-above private apartments. The first-floor retail stores shall be modular in construction so there could be

1. three stores with a 20′ front,

2. two stores with a 30′ front, or

3. one store with a 20′ front and a second store with a 40′ front

Each store will have a separate and controllable access to its portion of the basement. Each store and private apartment shall have its own entrance and rear exit, heating, ventilation and air conditioning system, metered electrical power, communications, natural gas, water, and sewage. A common elevator shall be provided for both freight and passenger use. There shall be two stairways for each floor of apartments. Those stairways shall be on opposite ends of the building for fire-exit purposes. A hallway shall be provided for each private apartment.

The second structure shall contain only offices and shall be modular in construction. Each office shall have its own entrance, heating, ventilation and air conditioning system, metered electrical power, communications, natural gas, water, and sewage. There shall be two stairways for each floor of offices. Those stairways shall be on opposite ends of the building for fire-exit purposes. A hallway shall be provided for access to all offices on the second floor and above. A common elevator shall be provided for both freight and passenger use.

The third structure shall contain warehoused materials that are separated into either two or three modular sections. Each warehouse section shall have its own entrance and rear exit, heating, ventilation and air conditioning system, metered electrical power, communications, natural gas, water, and sewage. Each loading-dock entrance shall be large enough to accept materials that are at least 10′ wide and 6′ high. If additional floors are included in the design, then separate elevators shall be provided for freight and passenger use. (The size of the freight elevator and the floor loading characteristics shall be determined as a part of the design process.)

Separate elevators shall be provided for freight and passenger use. There shall be two stairways for each floor of the warehouses. Those stairways shall be on opposite ends of the building for fire-exit purposes. A hallway shall be provided for each warehouse.

All three buildings will be provided with electrical power, communications, gas, water, and sewage connections.

BIOMEDICAL AND BIOTECHNOLOGY

CHEMICAL

CIVIL

System Definition

Identify one or more abutting properties that can be surveyed and designed for multiple use as housing and industrial subdivisions, with perhaps a small airport. Survey the identified property(s). Prepare plot plans and topographic plans. Determine existing zoning constraints such as setbacks and permitted usage of the land. Establish vehicular access that coordinates with adjacent properties.

System Characteristics

Devise a layout of the property(s) subdivided into

1. wildlife refuges and wetlands,
2. agricultural,
3. residential,
4. commercial, and
5. industrial.

Consider the need for buffer zones, roadways, bridges, drainage, electrical power, natural gas, and sewage installations that take advantage of the land contours, trees, rock outcroppings, and waterways. Describe any constraints regarding future structure location, size, style, and exterior finish.

COMMUNICATIONS

System Definition
System Characteristics

COMPUTER

System Definition

Devise a portable computer that shall be compatible with a variety of accessory and communication devices. It is to be easy to transport, set up, use, and store and shall operate and/or be chargeable from a variety of external power sources. It shall be operable in an environment suitable to an ungloved operator operating in a safe manner. The processing abilities and cost of the total package and its accessories shall be competitive to those expected to be marketed within the same time frame as this portable computer. That time frame is to be within eighteen months from concept initiation. The useful lifetime of this computer and its accessories shall be at least three years.

System Characteristics

The portable computer shall consist of a data processor and controller, keyboard, mouse or equivalent pointing device, monitor, 8.9 cm (3.5″) floppy disk drive, and power supply. Connections shall be provided for its use with an external monitor, CD-ROM drive, printer, optical character reader, and communications devices. An optional carrying case is to be either selected from one already in stock or designed as a new accessory. When the computer is removed from the carrying case, its maximum base dimensions shall permit its use on a small portion of a standard desk top. It shall contain no exposed sharp edges. There shall be no access by the user to internal electrical power.

The data processor and controller shall utilize standard components. These components are to be compatible with the manufacturing and repair-parts market availability during the four and one-half years from computer conception through its expected lifetime.

The keyboard and mouse or equivalent pointing device shall be designed ergonomically to minimize long-term physical stress such as carpal tunnel syndrome. The monitor shall be designed to minimize eye strain. The locations of the keyboard, mouse or equivalent pointing device, monitor, and any other related controls shall be compatible with each other and with the intended user.

The 8.9 cm (3.5″) floppy disk drive shall be able to receive disks formatted for either Macintosh or PC applications. The portable computer shall be able to both read from and write to disks contained within the floppy disk drive. The drive shall automatically protect itself when not in use.

The external power source shall be either 115 V, 50 to 60 Hz, or a car cigarette lighter socket. An internal power supply, possibly a combination line rectifier and battery, shall provide internal power, line surge filtering, and external power regulation. A combined audible and visual signal shall be provided to warn the user that the energy being supplied from the internal power source is within a few minutes of being unable to permit computer operation.

Connections for use with an external monitor, CD-ROM drive, printer, optical character reader, and communications devices shall meet IEEE Standard #XXX.X. Communications devices shall include infrared computer-to-computer links and hard-wired standard telephone jacks. Any required modem shall be contained within the portable computer.

The optional carrying case should contain at least one pocket that can store up to ten disks and another pocket that can store a pad of paper that is no more than 22 cm by 28 cm (8.5″ by 11″). When closed, the case shall resist the penetration of rain and snow. When stored, the case shall protect the computer from damage due to industrial-level shock, vibration, temperature, and humidity conditions.

ELECTRICAL AND ELECTRONIC

System Definition

System Characteristics

Describe system concept—attributes, functions, and features

Attributes—comfort, portability, durability

Functions—to be determined later

Features—safety, cost, and speed

Provide bounds and/or goals for system performance

    What the system must do; not what it must be

Indicate design boundaries (constraints), both technical and economic
Identify system and subsystem functions

ENVIRONMENTAL

System Definition
System Characteristics

HAZARDOUS WASTE

INDUSTRIAL

MANUFACTURING

MARINE AND MARINE SCIENCE

MATERIALS

MECHANICAL

NUCLEAR

OPTICS

System Definition
System Characteristics

SYSTEMS

System Definition

    Devise a system that can sense, display, and activate existing and planned remote analog and digital devices. The system shall be designed to

1. sense and store information in a digital format,

2. transmit digital information to one or more remote locations, and

3. receive, store, and implement digital instructions.

The system shall consist of a family of compatible sensors, transducers, data entry devices, controllers, recorders, displays, transmit/receive devices, and their necessary power subsystems.

Human factors considerations shall be an important part of the design. Problems related to operation during quiet times and when the operators are being stressed by either or both the system and the environment are to be considered.

Each device within the system shall be designed for both commercial and industrial purposes. The format and data content shall be programmable so as to lend itself to a variety of applications.

Parameters to be monitored and controlled can be as simple as temperature, humidity, and water levels. They can also be as sophisticated as identifying out-of-specification parameters of a process or a security system so alarms can be automatically transmitted to other locations.

Initiate a design that approximates the system characteristics described below. Then perform a marketing study to determine competitive existing and expected systems. Also determine the preferences of potential customers regarding flexibility, cost, and maintenance and repair expectations. Estimate the size of each market studied.

System Characteristics

The sensors and transducers shall have the ability to measure the parameters for which they were designed. They shall provide their information in a digital format to be determined later. The sensors shall provide that information to a controller when they are interrogated. The digital communication technique will be determined as a part of the design.

Parameters to be monitored and controlled shall include temperature, humidity, wind speed and direction, and any other parameters that can be presented in a digital format. Controls, such as on/off functions, shall also be available. The range of parameters to be monitored and controlled shall depend upon a study of the potential market.

The data entry devices shall be keyed by a human being or by previously recorded data that are entered based upon clock time. This information will be stored by one or more of the applicable recorders.

The controllers are the heart of the system. Each controller shall perform at least the following functions:

1. Exercise control over a minimum of four parametric functions and a maximum that is to be determined during a market study.
2. Provide power to all active devices.
3. Provide timing signals to all connected devices.
4. Transmit and receive data between units as programmed.
5. Provide data location, data value, and time-clock information to the connected recorders and to backup recorders.
6. Connect to a keyboard.
7. Contain computational circuitry that can provide data for each connected display in the format described for that display.
8. Accept preprogrammed ROM circuitry designed for a specific application that includes the quantity and type of input, output, and transmit/receive devices and information that describes those given parameters that are out of specification. The preprogrammed ROM parameters shall be retained within the controller in such a way that an authorized operator can modify them. The controller will transmit the identification of the authorized operator and the time of parameter modification to the appropriate recorder and its backup recorder.
9. Identify out-of-specification parameters, select one or more preidentified telephone numbers from a recorder, activate one or more calls until a human being responds, and verify that the person who has received that emergency notification has responded with a correct identification code.

The recorders shall store all data delivered to them from either the controllers or the data entry devices. That data shall be stored in a common format for each sensor, transducer, and data entry device. The data shall be coded with respect to time of information arrival.

The displays shall consist of at least two types—one small portable display and one desktop display.

1. The smaller display shall provide, as a minimum, eight alphanumeric characters per row for two rows. The first row shall indicate the source and type of data; the second row shall indicate the value of that data. The smaller display shall also provide for alphanumeric keyboard entry.

2. The larger display shall provide, as a minimum, the ability to present information in both alphanumeric and graphical format. The graphical format shall provide a vertical axis labelled with the parameter numeric data, including the appropriate units and a horizontal axis labelled as a "time" axis with the appropriate units. The vertical axis shall be convertible from metric units to U.S. common units upon command. The larger display shall also provide for alphanumeric keyboard entry.

The presentation of both displays shall be controllable by a properly identified operator using a connected data entry device.

The transmit/receive devices shall contain a slow-speed modem and shall operate via existing analog and digital

(a)   telephone lines and

(b)   vehicular communications systems.

In the field where there may be no controller, the transmit/receive device shall have the ability to interrogate existing sensors and transducers. Whenever a controller is present, the devices shall be slaved to that controller. It may be desirable to develop two or more separate transmit/receive devices, depending upon information gathered during a study of the potential market.

The power subsystem shall consist of at least three units:

1. A power-storage unit that can remain in the field and can be charged by solar and/or wind power.

2. A vehicular power-supply unit that can operate from the 12 Vdc of that vehicle or from conversion from that vehicle's 12 Vdc to whatever voltage or voltages are required by the equipment being powered. The power supply shall have the ability to be connected to a receptacle such as a cigarette lighter or other receptacle associated with vehicular communications.

3. A power supply that is operated from a 115/230 V, 50 Hz to 60 Hz source of electrical power.

APPENDIX H
Student Activity Center
As-Built Plans

GENERAL

The As-Built Plans for the Student Activity Center consist of eleven drawings:

Topographical Plan of Property
Site Plan
Foundation Plan
Basement Plan
First Floor Plan
Attic Crawl Space Plan
Cross Section
Long Section
End Elevations
Long Elevations
Door and Window Schedule

Their content is given below.

TOPOGRAPHICAL PLAN OF PROPERTY

Property Line Dimensions and Directions (as magnetic or GPS bearings)
Final Grades and Lot Drainage

SITE PLAN
(SHOWING STUDENT ACTIVITY CENTER LOCATION)

Utility Connections
Water
Sewer
Natural Gas
Electricity
Telephone
Cable Television

FOUNDATION PLAN

Floor Drain Piping
Foundation Wall Footings
Chimney Footings
Lally Column Bases
Furnace Base

BASEMENT PLAN

Floor Drains
Basement Floor
Utilities Entrances and Exit (water and sewer)
Stairway
Furnace Location
Water Heater Location
Lavatory Content and Piping Locations
Partitions
Door(s)

FIRST FLOOR PLAN

Stairway
Windows and Doors Locations
Partitions
Heat and Duct Layout and Locations
Lavatory Content and Piping Locations
Kitchen Content, Layout, and Locations
Storage Shelves Locations

ATTIC CRAWL-SPACE PLAN

Entrance Hole
Ventilation Louver Locations

CROSS SECTION (END VIEW)

Footings and Applicable Walls
Chimney
Trusses

Ridge Vent

Doorways where applicable

LONG SECTION (SIDE VIEW)

Footings and Applicable Walls

Floors

Stairways

Rafters

Chimney

END ELEVATIONS

Entrance Ways

Windows

Chimney

Typical Exterior Finish

LONG ELEVATIONS

Entrance Ways

Windows

Chimney

Typical Exterior Finish

DOOR AND WINDOW SCHEDULE

Type

Quantity

Location

Glossary of Terms

GENERAL

This glossary contains a brief definition of both technical and nontechnical terms. These definitions focus on how the words are used in this text.

Access code. A software security code that is used to control access to a database.

Action item. Work identified during a meeting that requires immediate effort to correct or modify the budget, task, or schedule of performance.

Activity. A process such as searching, learning, doing, or writing, that involves the application of both a mental and a physical operation; an expenditure of work or energy is required. An action or pursuit that relates to the planning or performance of a project or program.

Actuals. The record of the expenditures of a project or organization. It is a record of the monies spent during a defined timeframe. *See also* estimates.

Administrators. Persons who monitor and control an overall project and who can visualize the final goal of that project. They judge performance and guide the project team.

Analysis. The gathering of data. The separating or breaking up of the whole into its parts so as to determine their nature, proportion, and function. The solving of problems. The process of logical reasoning that leads from stated premises to the conclusion concerning specific capabilities of equipment or software and its adequacy for a particular application. The process of listing all design requirements and the reduction of those requirements to a complete set of logically related performance specifications.

Applied research. From "search again." A careful, patient, systematic, diligent, and organized inquiry or examination in a field of knowledge. It is undertaken to establish facts or principles. It is a laborious and continued search after truth or information. Its goal is to convert an idea into a practical plan.

Archival documents. Documents retained for reference and historical purposes.

Bionics. An abbreviation for *biology-electronics*. It is the application of biological principles to the study and design of engineering systems that often involve animals and plants. It is also known as *nature analysis*.

Block diagram. A diagram of interconnected rectangles known as "blocks." The input to a block, the function(s) of that block, and the output from that block must be specified; each block represents a part of the product; the interconnections of various blocks indicate the relationships among the parts.

Budget. A plan that describes the types and amounts of authorized expenses for a specified period of time.

Brainstorming. The unrestrained offering of ideas and suggestions by all members of a group during a meeting that typically focuses on a predetermined range of items. During a brainstorming session, no one is allowed to comment with negative remarks. Instead they are encouraged to expand and enlarge on ideas as they are presented.

Byte. A unit of computer memory. Eight computer bits.

CPM. *See* Critical Path Method

Charge numbers. Numerics or alphanumerics used to uniquely identify a task or subtask.

Charges, time. *See* time charges.

Code, access. *See* access code.

Component.   A portion of a device that normally performs a useful function only when used within that device.

Conceivers.   *See* creators.

Concurrent engineering.   A systematic approach that integrates the initial idea for a project with its applied research, design, development, marketing, production, sales, and distribution. It leads to a more efficient and shorter design cycle while maintaining or improving product quality and reliability.

Conflict resolution.   The work of executive personnel to reduce or eliminate dissension, strife, and perhaps hostility among team members.

Construction.   *See* production.

Control(ling) of a project.   The act of seeing that the project proceeds according to a well-prepared plan.

Costs, direct.   *See* direct costs.

Costs, fixed.   *See* fixed costs.

Costs, indirect.   *See* indirect costs.

Costs, variable.   *See* variable costs.

Creativity.   The art of devising new ideas that may be ingenious, inventive, or innovative.

Creators.   Persons who have the skills of imagination and vision and can think about a problem and devise one or more solutions.

Critical Path Method (CPM).   A means for indicating which task sequence will have the greatest effect on schedule performance.

Database.   A collection of related information associated with a specific organization or project.

Database, common.   A collection of related information. It can be accessed by a variety of users and can be updated (modified with later information) as required.

Deliverables.   Tangible, verifiable products such as a feasibility study, a detailed design, or a working prototype. The conclusion of a project phase is generally marked by a review of deliverables (such as end-of-phase documents) and project status. Deliverables include both equipment and documents that are provided to a client.

Delphi (approach to selecting a problem solution).   Applies numeric measures to potential problem solutions.

Design.   Refers to the conversion of an idea into a plan for a product, process, or procedure by arranging the parts of a system via drawings or computer models so that the part will perform the desired function when converted to an actual product.

Design options.   Choices that are offered to a client or to the managers in order that they may participate in the selection process.

Desire.   Similar to a wish, it is a longing for something that promises enjoyment or satisfaction. A desire is not a need. A strong desire may eventually become a need; a weak desire may be considered a luxury.

Deterministic.   Each activity is determined by a sequence of events over which no one has control. A foundation must be constructed prior to building the walls, for example.

Development.   The activity of modifying a design, product, process, or procedure.

Device.   The portion of a subsystem that can perform an identifiable function either alone or with other parts of that subsystem.

Direct costs.   Labor and material costs that are part of a cost package.

Distribution.   The process of transporting completed products to warehouses, customers, or retail outlets.

Document, trial.   *See* trial document.

Documentation.   The records developed to describe a planned or performed effort. It includes words, figures, drawings, tables, and/or graphs.

Documents, archival.   *See* archival documents.

Ergonomics.   *See* human factors.

Estimates.   The record of the planned expenditures of a project or organization.

Evaluation.   Determining the accuracy with which alternative designs fulfill performance requirements for operation, manufacture, and sales before the final design is selected.

Field.   A group of items that contains information regarding a person, place, or thing.

Fixed costs.   Costs whose amount can be determined and will remain unchanged for an extended time. Costs that are independent of the amount of engineering or production.

Form.   The shape, outline, or solution that results from the definition of a function or system function. "Form follows function."

Function.   The act or operation expected of a person or thing. The ability of an individual, process, device, or equipment to perform in a given manner. It is sometimes provided as a system function.

Gantt chart.   A technique for presenting related schedule information in the form of a graph.

Goal.   An object of desire or intent. That which a person sets as an objective to be reached or accomplished. The end or final purpose. An ambition, aim, mark, objective, or target.

GPS (Global Positioning System). A satellite-based system that precisely locates (within a few meters) a position on our earth.

Ground rules.   Rules that govern all conduct, procedures, and aspects of the proposed project output. They are typically established during the latter part of the Conception Phase or during the Study Phase.

Hierarchy.   A group of persons or things that are arranged in a predetermined order of rank, grade, or class.

Human factors.   The application of knowledge regarding human beings through the design of products or procedures.

Idea.   A concept first formulated in the mind of one or more persons. It may be a concept that can be designed or developed immediately, or it may require further study.

Indirect costs.   Costs that are not directly related to the design or production of an item or service. All costs that are not direct costs.

Interface.   The mechanical and electrical boundary for two or more equipments. Cables, connectors, transmitting and receiving circuits, signal-line descriptions, timing, and control information are included.

Item.   A piece of information that is to be stored for later processing.

Item, action.   *See* action item.

Iterative.   Said or performed repeatedly until the result or product is acceptable.

Just-in-time.   The timing determined so that materials will be delivered or work will be performed no sooner than is necessary to meet schedule requirements.

Laborers.   Persons who work primarily with their hands on physical activities.

Letter contract.   A brief document that contains both legal and technical information. It defines the work to be accomplished, its schedule, and its price. It is used where the work is simple in nature and does not require a lengthy contract with applicable attachments.

Letter of transmittal.   A letter that contains both legal and technical information. It accompanies a contract of significant length and refers to that contract and its applicable attachments.

Life cycle (of a project plan).   Starts with the initial idea for a project and continues through applied research, design, development, marketing, production, sales distribution, and, in some cases, maintenance, disassembly, and salvage or disposal.

Logistics.   From the beginning to the end of the supply chain, the procurement of raw materials and components. The acquisition, storage, movement, and distribution of subassemblies and finished products.

Luxury.   The best and most costly thing that offers the most physical, intellectual, or emotional comfort, pleasure, and satisfaction.

Make-or-buy decision.   The option that relates to that portion of a design that will be fabricated within the organization (make) or will be purchased from vendors (buy).

Management, bottom-up.   An organized structure where the lowest level of personnel offers suggestions that are gathered, coordinated, and acted upon by the next higher level of personnel.

Management by exception.   The monitoring of an activity or project. Control is exercised only when a person or group does not meet its assigned goals.

Management, top-down.   An organized structure where the highest ranking leader is personally involved in directing all other personnel via verbal orders and written directives.

Manager.   One who holds an administrative, executive, or managerial position. Normally one who directs the performance of a supervisor.

Market Research.   The process of determining the need for a new product, process, or procedure, and the amount of money that potential customers are willing to invest. Market research must also determine if it will sell in sufficient quantities to meet the total project costs as well as earn a profit.

Marketing.   The process of selling a given product, and determining its selling price and selling potential.

Model.   An abstraction and simplification of an actual event or function. It simulates the actual system, or portion of that system, in an effective and efficient manner. A representation of selected features of a system. It may be a physical scale model of a proposed structure, it may be a defining set of equations, or it may be an operating prototype of a product.

Monitor.   To advise, warn, or caution a person, activity, or project.

Nature Analysis.   *See* bionics.

Need.   Something that is considered necessary or essential. There is a hierarchy of needs. (*See* page 23 of the text.)

Objective.   Describes the work to be performed. It is included within a Specification document or in a separate Project Task Description document. It is analogous to a road map in that it indicates the direction to be pursued without restricting the designer to a specific result.

Optimum.   The best or most favorable degree, condition, or amount of a design that is not necessarily the best for each individual item, but is the best for the overall design. (Many common design problems have multiple local optima, as well as false optima, that make conventional optimization schemes risky.) *Optimum* is a technical term that infers a mathematical, quantitative context.

Organization.   Any unified, consolidated body of persons working together for a specific purpose.

Organize.   To plan in a methodical manner. To bring about an orderly disposition of individuals, units, or elements. To assemble resources and prepare for their use.

Part.   *See* component.

Perform.   To do what is required to achieve a desired goal. To fulfill or complete a process, product, or procedure to a successful conclusion.

Performance reviews, contract.   A document, used in conjunction with a face-to-face interview, that evaluates the activities of those individuals working on a given contract. It offers both positive evaluation and constructive criticism.

Performance reviews, personnel.   A document, used in conjunction with a face-to-face interview, that evaluates the activities and behavior of an individual who works with other persons. It offers both positive evaluation and constructive criticism.

Performers.   Persons who enjoy working with details, who are comfortable with developing and implementing the ideas of others, and receive satisfaction from coordinating a variety of activities.

PERT chart.   A technique for presenting related schedule information in the form of a graph. The precedence of each activity is indicated by means of linking arrows.

Plan.   The document that describes the conversion of ideas into a product, process, or procedure. It typically consists of budgets, tasks, and schedules accompanied by a description of what is to be accomplished.

Planners.   Persons who prepare the details of a project and analyze estimates.

Process.   A series of actions that leads to a result.

Production.   The act of fabricating numerous, identical copies of an item.

Program.   A group of related projects.

Project.   An organized effort to accomplish something useful or attain a useful end result. It is sometimes referred to as a plan, venture, or enterprise.

Purpose.   That which one proposes to accomplish or perform. The reason that one wants to accomplish something. An aim, intention, design, or plan.

Rationale.   A statement of causes or motives. An explanation, justification, or reason.

Record.   The name of a document that contains information worth developing, distributing, and maintaining. It is also the name of a portion of a computer file that contains more than one field of data.

Record, structured.   A set of categories of related data that is identical in content, arrangement, and length containing more than one field of data.

Record, unstructured.   A set of categories of data related to one another. Every item within each field may be different in content, arrangement, and length that contains more than one field of data.

Recycling.   Stripping into pieces. Also known as *deManufacturing* (tearing apart). Recycling should be planned during the Conception Phase.

Return on investment.   The percent of money earned based upon the amount of money loaned.

Sales.   Identifying and contacting customers to convince them to purchase an item.

Slack time.   The extra time available in a schedule to allow for unplanned changes that may occur during the performance of one or more related tasks.

Specification.   A document that describes the work to be accomplished, including the conditions (requirements) that must be fulfilled in order to complete a task or project.

Spreadsheet.   The presentation of an organized arrangement of data in matrix (tabular) form.

Stakeholder.   A person or organization that has a financial investment (stake) in a project or product. A person or organization that monitors its performance to ensure that their investment is efficiently and successfully applied.

Structured record.   *See* record, structured.

Subsystem.   A portion of a system that performs one or more functions that can, alone or in a system, operate separately.

Synthesis.   The putting together of separate parts or elements so as to form a connected whole. It is a qualitative process and is the opposite of analysis.

Synthesize.   The bringing together of separate parts into a coordinated unit.

System.   A combination of things that actively work together to serve a common purpose. They are composed of combinations of parts, people, and procedures.

Systematic approach.   An orderly and methodical way of working when conceiving, studying, designing, and implementing a project.

Task.   An undertaking or portion of work assigned to be accomplished. Sometimes known as a job.

Task description.   A record that describes the name and objective of a task, the relation of that task to other tasks, a description of the work to be performed, the specification(s) relating to any materials designated, an explanation of the expected quality of work, and a schedule for the performance of that task.

Time charges.   Information related to the number of hours that each individual has worked on a given task or project.

Trade-off.   The exchange of one benefit or advantage to gain another benefit or advantage that is considered to be more desirable.

Trial document.   A document devised to resolve an uncertainty. It is not considered an official document until the appropriate managers review and approve it.

Unstructured record.   *See* record, unstructured.

Variable costs.   Costs that will change depending upon the production level of an item or the amount of service provided to a customer.

Warranty.   A document that indicates the extent to which the provider of parts and procedures is responsible for their performance.

Bibliography

Allen, M. S. *Morphological Creativity: The Miracle of Your Hidden Brain Power.* Upper Saddle River, NJ: Prentice-Hall, 1962.

American Institute of Architects (AIA) and Associated General Contractors of America (AGCA). *Abbreviated Form of Agreement between Owner and Contractor with Instruction Sheet.* AIA Document 107. 9th ed. Washington DC: American Institute of Architects (AIA) and Associated General Contractors of America (AGCA), 1987.

American Institute of Architects (AIA) and Associated General Contractors of America (AGCA). *Recommended Guide for Competitive Bidding Procedures and Contract Awards for Building Construction.* Washington, DC: American Institute of Architects (AIA) and Associated General Contractors of America (AGCA), 1980.

Archer, L. B. *Systematic Method for Designers.* Chichester, U.K.: Wiley, 1984.

Archibald, R. D. *Managing High-Technology Programs and Projects.* New York: John Wiley & Sons, 1992. ISBN 0-471-51327-X.

Badiru, A. B. *Project Management Tools for Engineering and Management Professionals.* Norcross, GA: Institute of Industrial Engineers, 1991. ISBN 0-89806-114-8.

Badiru, A. B., and Whitehouse, G. E. *Computer Tools, Models, and Techniques for Project Management.* Blue Ridge Summit, PA: TAB Books, 1989. ISBN 0-88906-3200-X.

Banios, E. W. "An Engineering Practices Course." *IEEE Transactions on Education,* 35, no. 4 (November 1992): 286–293.

Boff, K. R., and Lincoln, J. E., eds. *Engineering Data Compendium: Human Perception and Performance.* Wright-Patterson Air Force Base, OH: Henry G. Armstrong Aerospace Medical Laboratory, 1988.

Bransford, J. D., and Stein, B. S. *The Ideal Problem Solver.* New York: W. H. Freeman, 1993. ISBN 0-7167-2204-6.

Brown, J. *Value Engineering—A Blueprint.* New York: Industrial Press Inc., 1992.

Callahan, M. T., Quackenbush, D. J., and Rowings, J. E. *Construction Project Scheduling.* New York: McGraw-Hill, 1992. ISBN 0-07-009701-1.

Carrithers, W. M., and Weinwurm, E. H. *Business Information and Accounting Systems.* Columbus, OH: Charles E. Merrill, 1967. ISBN 675-09749-5.

Chase, W. P. *Management of System Engineering.* New York: John Wiley & Sons, 1974. ISBN 0-89874-682-5.

Clark, C. H. *Brainstorming: How to Create Successful Ideas.* North Hollywood, CA: Wilshire Book Company, 1958.

Clawson, R. H. *Value Engineering for Management.* Princeton: Auerbach Publishers, Inc., 1970.

Connor, P. E., and Lake, L. K. *Managing Organizational Change.* Westport, CT: Praeger, 1994. ISBN 0-275-94652-5.

Construction Specifications Institute, P.O. Box 85080, Richmond, VA 23285-4236, 1998.

Cross, N. *Engineering Design Methods.* Chichester, U.K.: John Wiley & Sons, 1944.

Dell'Isola, A. J. *Value Engineering in the Construction Industry.* New York: Van Nostrand Reinhold, 1982.

Dinsmore, P. C. *Human Factors in Project Management.* New York: American Management, 1990. ISBN 0-8144-5003-2.

Duncan, W. R. *A Guide to the Project Management Body of Knowledge.* Project Management Institute, 130 South State Road, Upper Darby, PA 19082, 1996.

East, E. W., and Kirby, J. G. *A Guide to Computerized Project Scheduling.* New York: Van Nostrand, Reinhold Co., 1990. ISBN 0-442-23802-9.

Elsayed, E. A., and Boucher, T. O. *Analysis and Control of Production Systems.* Upper Saddle River, NJ: Prentice Hall, 1985.

Farkas, L. L. *Management of Technical Field Operations.* New York: McGraw-Hill, 1970. L.O.C. 76-107289.

Fasal, J. H. *Practical Value Analysis Methods.* New York: Hayden Book Co., 1972.

Fisk, E. R. *Construction Project Administration.* 4th ed. Englewood Cliffs, NJ: Prentice Hall, 1992. ISBN 0-13-174137-3.

Francks, P. L., Testa, S. M., and Winegardner, D. L. *Principles of Technical Consulting and Project Management.* Chelsea, MI: Lewis Publishers, 1992.

Goode, H. H., and Machol, R. E. *System Engineering.* New York: McGraw-Hill, 1957.

Graham, N. *Introduction to Computer Science: A Structural Approach.* St. Paul, MN: West Publishing Company, 1982. ISBN 0-314-64968-9.

Graham, R. J. *Project Management: Combining Technical and Behavioral Approaches for Effective Implementation.* New York: Van Nostrand Reinhold Co., 1985. ISBN 0-442-23018-4.

Heyne, P. *The Economic Way of Thinking.* 6th ed. Toronto: Macmillan Publishing Company, 1991. ISBN 0-02-354181-4.

Horgan, M. O'C., and Roulston, F. R. *The Foundations of Engineering Contracts*. 1989. ISBN 0-419-14940-6.

Hosny, O. A., Benjamin, C. O., and Omurtag, Y. "An AI-based Decision Support System for Planning Housing Construction Projects." *Journal of Engineering Technology,* Fall 1994: 12–19.

IEEE *Standard Dictionary of Electrical & Electronics Terms*. New York: Wiley Interscience, 1984.

Isaak, J. "Standards for the Information Infrastructure." *IEEE Standards Bearer,* July 1994: 10.

Jones, J. C. *A Method of Systematic Design*. Chichester, U.K.: Wiley, 1984.

Kaimann, R. A. *Structured Information Files*. Los Angeles: Melville Publishing Company, 1973. ISBN 0-471-45483-4.

Kaplan, G. "Manufacturing, Á La Carte." *IEEE Spectrum,* September 1993: 24–85.

Karwowski, W., ed. *International Encyclopedia of Ergonomics & Human Factors*. Taylor & Francis, 2000. ISBN 0-7484-0847-9.

Kim, S. H. *Essence of Creativity*. New York: Oxford University Press, 1990. ISBN 0-19-506017-2.

Lewis, J. P. *Project Planning, Scheduling, and Control*. Chicago: Probus Publishing Company, 1991. L.O.C. 91-60939. ISBN 0-9631886-0-7.

Lumsdaine, E., and Lumsdaine, M. *Creative Problem Solving*. New York: McGraw-Hill, 1990. ISBN 0-07-039087-8.

Mager, R. F. *The New Mager Six-Pack,* Belmont, CA: Lake Publishers, 1996. ISBN 1-879-61801-X.

March, L. J. *The Logic of Design*. Chichester, U.K.: Wiley, 1984.

Metedith, J. R., and Mantel, S. J. *Project Management: A Managerial Approach*. New York: John Wiley and Sons, 1995. ISBN 0-471-01626-8.

Miles, L. D. *Techniques of Value Analysis and Engineering*. New York: McGraw-Hill Book Company, 1972.

Mudge, A. E. *Value Engineering: A Systematic Approach*. McGraw Hill, New York: 1971.

Muller, E. J., Fausett, J. G., and Grau, P. A. *Architectural Drawing and Light Construction*. 5th ed. Upper Saddle River, NJ: Prentice Hall, 1998. ISBN 0-13-520529-8.

Nadler, G. *Work Design: A Systems Concept*. Homewood, IL: Richard D. Irwin, 1970. L.O.C. 70-110417.

Nicholas, J. M. *Managing Business and Engineering Projects*. Englewood Cliffs, NJ: Prentice Hall, 1990. ISBN 0-13-551854-7.

Oberlender, G. D. *Project Management for Engineering and Construction*. New York: McGraw-Hill, 1993. ISBN 0-07-048150-4.

Osborn, A. *Applied Imagination—The Principles and Problems of Creative Problem Solving*. 3rd ed. New York: Charles Scribner's Sons, 1963.

Pahl, G., and Beitz, W. *Engineering Design*. London: Design Council, 1984.

Priest, J. W., Bodensteiner, W., and Muir, N. K. "A Survey of Educational and Training Needs for Transition of a Product from Development to Manufacturing." *IEEE Transactions on Education,* 37, no. 1 (February 1994): 13–22.

Pugh, S. *Total Design: Integrated Methods for Successful Product Engineering*. Reading, MA: Addison-Wesley Publishing Company, 1993.

Raferty, J. *Risk Analysis and Project Management*. London: E & FN, 1994. ISBN 0-419-18420-1.

Rechtin, E. "The Synthesis of Complex Systems." *IEEE Spectrum,* July 1997: 51–55.

Reutlinger, S. *Techniques for Project Appraisal under Uncertainty*. Baltimore: Johns Hopkins Press, 1970. ISBN 0-8018-1154-6.

Roze, M., and Maxwell, S. *Technical Communications in the Age of the Internet*. Upper Saddle River, NJ: Prentice Hall, 2002. ISBN 0-13-020574-5.

Salvendy, G. *Handbook of Human Factors and Ergonomics,* 2nd ed. New York: John Wiley & Sons, 1997. ISBN 0-471-11690-4.

Samaris, T. T., and Czerwinski, F. L. *Fundamentals of Configuration Management*. New York: Wiley Interscience, 1971. ISBN 0-471-75100-6.

Sanders, M. S., and McCormick, E. J. *Human Factors in Engineering and Design*. 7th ed. New York: McGraw-Hill, 1992. ISBN 0-07-054901-X.

Shepherd, J. M. *How to Contract the Building of Your New Home*. 3rd ed. Williamsburg, VA: Shepherd Publishers, 1991. ISBN 0-9607308-7-7.

Shim, J. K., and Siegel, J. G. *Theory and Problems of Managerial Accounting*. Schaum's Outline Series. New York: McGraw-Hill, 1984. ISBN 0-07-057305-0.

Smith, R. P. "Teaching Design for Assembly Using Product Disassembly." *IEEE Transactions on Education, 41,* no. 1 (February 1998): 50–53.

Steiner, C. J. "Educating for Innovation and Management: The Engineering Educators' Dilemma." *IEEE Transactions on Education, 41,* no. 1 (February 1998): 1–7.

Stolarski, T. A. *Analytical Techniques*. Elsevier Science B.V., 1995 Edition.

Tapper, J. *Evaluating Project Management Software: How, Why, and What*. jtapper@coe.neu.edu, 1998.

Texas Instruments, Inc. *Working with Suppliers*. TI-29503.

Turner, W. C., Mize, J. H., and Case, K. E. *Introduction to Industrial and Systems Engineering*. Upper Saddle River, N.J.: Prentice Hall, 1987.

Van Cott, H. P., and Kinkade, R. G. *Human Engineering Guide to Equipment Design*. Washington, DC: American Institutes for Research, Superintendent of Documents, U.S. Government Printing Office, 1972. (Includes an extensive bibliography.)

Verein Deutscher Ingenieure (VDI). *VDI-Guidelines*. VDI 2221, *Systematic Approach to the Design of Technical Systems and Products*. Germany Professional Engineers' body.

von Oech, R. *A Whack on the Side of the Head*. New York: Warner Books, 1993. ISBN 0-446-77808-7.

Wilde, D. J. *Globally Optimal Design*. New York: John Wiley & Sons, 1978.

Index